ART AND SCIENCE IN BREEDING
Creating Better Chickens

Art and Science in Breeding
Creating Better Chickens

MARGARET E. DERRY

UNIVERSITY OF TORONTO PRESS
Toronto Buffalo London

© University of Toronto Press 2012
Toronto Buffalo London
www.utppublishing.com
Printed in Canada

ISBN 978-1-4426-4395-6

Printed on acid-free, 100% post-consumer recycled paper with vegetable-based inks.

Library and Archives Canada Cataloguing in Publication

Derry, Margaret E. (Margaret Elsinor), 1945–
Art and science in breeding : creating better chickens / Margaret E. Derry.

Includes bibliographical references and index.
ISBN 978-1-4426-4395-6

1. Chickens – Breeding – Canada – History.   2. Chickens – Breeding – United States – History.   3. Chickens – Genetic engineering – Canada – History.   4. Chickens – Genetic engineering – United States – History.   5. Chicken industry – Canada – History.   6. Chicken industry – United States – History.   I. Title.

SF488.C3D47 2012   636.500971   C2011-907815-5

This book has been published with the help of a grant from the Canadian Federation for the Humanities and Social Sciences, through the Aid to Scholarly Publications Program, using funds provided by the Social Sciences and Humanities Research Council of Canada.

University of Toronto Press acknowledges the financial assistance to its publishing program of the Canada Council for the Arts and the Ontario Arts Council.

 Canada Council for the Arts Conseil des Arts du Canada   ONTARIO ARTS COUNCIL
CONSEIL DES ARTS DE L'ONTARIO

University of Toronto Press acknowledges the financial support of the Government of Canada through the Canada Book Fund for its publishing activities.

Contents

Acknowledgments vii

Introduction 3

1 Historical Background: Chickens, Those Lowly Creatures 11

2 From Barnyard Scavenger on North American Farms to Bird of Beauty and Use 33

3 The Development of Agricultural Genetics in Relation to North American Chicken Breeding 74

4 Breeding for Eggs in North America: Conflict between Science and Craft 97

5 The 'Scientizing' of Breeding in the North American Egg Industry 128

6 North American Chicken Breeding and the Rise of the Broiler Industry 154

7 Epilogue: Trends in Chicken Breeding after 1950 178

Conclusion 202

Notes 215

Bibliography 253

Index 275

Acknowledgments

I had wonderful help writing this book, and I take pleasure now in thanking the people who provided it. Douglas W. Dodds, former chief executive officer of Schneider's Foods and currently chief strategy officer for Maple Leaf Foods, answered my many questions about how the modern poultry breeding industry functioned with both the processor/integrator and the grower sectors of the chicken industry. He introduced me to the supply management system that existed in Canada. He also put me in touch with geneticists connected with large chicken-breeding companies.

John Hardiman, animal geneticist and vice-president of research and development for Cobb-Vantress, Inc., gave me a good deal of information about the culture of chicken breeding and its interface with large integrators like Tyson Foods, which owns Cobb-Vantress. Dominic Elflick, animal geneticist and director of research and development for the international company Aviagen, told me much about how breeding interfaced with both integrators and growers in that country. Ken Laughlin, a wildlife biologist and geneticist located in Britain and group vice-president for policy and strategy for Aviagen, provided much information on the British background, and also supplied me with documents that I would not have been able to access in any other way.

I spoke to James Wilton, a professor of animal genetics at the University of Guelph, who explained more about how classical rather than molecular genetics underlay most research on practical animal breeding. Robert S. Gowe, a trained geneticist hired by the Canadian government who served as director of the Animal Research Centre, Research Branch, of Agriculture Canada for many years, explained

Canadian chicken-breeder attitudes to both the Record of Performance and to new quantitative genetic approaches of the 1950s and 1960s. Gowe would later be employed by the Shaver Poultry Breeding Farm. The breeder who started this world-wide company, Donald McQueen Shaver, also agreed to talk to me and to answer my many questions about how breeding companies operated in the 1970s and 80s.

Lyle Vanclief, former minister of agriculture for the Canadian federal government, spoke to me about supply management and quota systems as they pertained to both the Canadian egg and broiler industries.

When I had completed a rough draft of this manuscript, Terry Crowley, then chair of the History Department of the University of Guelph, very kindly agreed to read it. Because of the fractured nature of the information in the book, which arose from that illusive cleavage endemic to the story, I needed direction. He provided this, allowing me to see more clearly how to link themes together. The advice of a colleague can be invaluable in the writing of any history. I was fortunate to get wonderful advice on how to improve the book from a number of other readers from various academic institutions after the manuscript had reached a more advanced stage. My special thanks to Bert Theunissen, historian of science at the University of Utrecht in the Netherlands, for his careful reading and commentary of the book.

Special thanks to Len Husband, University of Toronto Press, for all his help. I am grateful as well for the wonderful comments made by the unknown reviewers. Thanks also to John St James for his work on the manuscript and to Frances Mundy for her help in guiding this book through to completion.

ART AND SCIENCE IN BREEDING
Creating Better Chickens

Introduction

This book, *Art and Science in Breeding*, focuses on patterns in historical poultry (fundamentally chicken) breeding and the organization of breeding in the United States and Canada between 1850 and roughly 1960. Its primary objective is to study how developing science (in this case genetics) interrelated with traditional practices followed on farms. I am interested, in other words, in the connection and nonconnection of agricultural livestock breeding with academic biology. I believe that addressing the farm/science linkage by dwelling as much on farm breeding and culture as on genetics can increase our understanding of how the interface worked. This is not a particularly common way for historians to approach the relationship of genetics to farming, but I think it is a worthwhile one to pursue. The focus helps elucidate, for example, the way practical (or what might be described as craft) breeding of livestock played a role in the development of genetics, and how government direction (instigated by both breeders and business concerns) shaped what was perceived to be scientific knowledge. The subject of historical chicken breeding in North America is especially useful in the study of such broad-ranging questions for two specific reasons. To begin with, chickens have become the most scientifically engineered of livestock. This trend developed first in North America and subsequently became a worldwide pattern. Genetics appeared to replace completely older, unscientific methods dating from the nineteenth century. In no other livestock industry has the so-called art of breeding been so overshadowed. But is this really what happened to poultry breeding? I attempt to answer that question. The North American chicken-breeding world also illustrates how regulations could parade as breeding methodology. I focus on organizational regulation because I believe it played a significant role in making it difficult to ascertain how craft breeding was actually practised. One of the chief difficulties in assessing how the art of breeding related to the science of breeding is the fact that it is often not clear what the art of breeding entailed.

The road to writing *Art and Science in Breeding* has been long and tortuous. It started when I began my animal-breeding research for a doctoral dissertation on the history of Canadian cattle breeding. When I revised the thesis for book publication (it became *Ontario's Cattle Kingdom*), an underlying question reared its head: namely, why did farm breeding seem to be completely separated from academic biology's approach to the problem? My subject, clearly concerned with biological phenomena, seemed to have nothing to do with the science of heredity. My second book, *Bred for Perfection* (which reviewed patterns of cattle, dog, and horse breeding in relation to marketing since 1800 in Britain, the United States, and Canada) drew me even farther into this strange paradox. Readers of *Bred for Perfection*, who came from a history-of-science background, also seemed puzzled by the implication that scientific thought had been largely irrelevant to domestic animal breeding. Increasingly, the scenario struck me as illogical. I determined to solve the mystery of the apparent dichotomy. In my research for *Horses in Society*, I made a greater attempt to pinpoint a connection between the two. In spite of my efforts, I failed to do so in any truly comprehensive way.

I decided then that I should make the farm/science linkage a central theme in a livestock-breeding study, and the result is this book, which uses chickens to study the topic. The global chicken population had become centred in North America by 1920, and so it is logical for a study of chicken breeding to focus on the area within this general time frame. Over 38 per cent of the world's chickens lived in the United States and Canada. The country with the next largest share was China, with nearly 16 per cent. Other single countries had far fewer; only two had as much as 6 per cent of the total. England and Wales, for example, contained less than 3 per cent of the world's chickens.[1]

Chapter 1 of *Art and Science in Breeding* establishes the background to North American craft-breeding attitudes between the late eighteenth and mid-nineteenth centuries. It looks at chicken history since domestication, and reviews what eighteenth- and early-nineteenth-century European livestock and poultry breeders had to say about the 'art' of breeding. The chapter also discusses the way European and North

American breeders organized breeding around certain constructs. Breeders argued that their work resulted in 'thoroughbred,' 'purebred,' or 'standardbred' animals. These definitions came to describe the culture of breeding, more than the methodology used in breeding, a point of some significance, as will be more evident later in the book. Chapter 2 describes how chickens on North American farms became birds of beauty and use rather than lowly creatures between 1850 and 1900. I review a variety of subjects in this chapter: the creation of the main American chicken breeds that became the basis of the modern international poultry industry, chicken-breeding principles in the United States and Canada, as well as the rise of an exhibition structure and its relationship to breeding principles. Certain important attitudes to breeding, endemic to either craft or genetically driven methodology, are also made clear; most notably, the critical division in opinions between breeder and producer or grower. Chapter 3 outlines the rise of agricultural genetics by 1940 in the United States. The 'science' of breeding was in fact based on many principles that governed the 'art' of breeding. Agricultural genetics, as it developed in American experiment stations, drew on craft-breeding knowledge, thereby making it difficult to distinguish originality in the geneticist approach. Fundamentally different outlooks lay at the heart of the matter.

Chapter 4 explores the art/science situation in chicken breeding after 1900, and in the process explains how a deeply divisive conflict arose. Chapter 5 looks at how chicken breeding became 'scientific' by describing North American hatchery-industry developments between 1920 and the 1940s. Hatchery men, who bought eggs from breeders and ran incubators that hatched the eggs, controlled what chickens would reach farmers. When breeding passed from individual craft breeders to companies with geneticists in control of breeding, fundamental changes occurred in the hatchery industry. Hatchery men no longer had a say in what type of chick they hatched. Government systems, which had bolstered traditional breeder methodology and organizational structures, collapsed and thereby fed into the situation. Chapter 6 describes how agricultural genetics made the modern broiler or meat industry possible. I look at breeding and the dynamics of a business structure that remained separate from breeding. Chapter 7 gives an overview of chicken-breeding developments after 1950, and assesses international corporate chicken breeding in the modern world. I use the papers of one breeding company to show how secrecy worked, how important globalization became to breeders, and how companies protect their breeding work. The chapter uses Canada's supply management system with respect to the broiler industry to show how important husbandry practices could be for breeder companies who faced increasing competition in the international industry.

This book contributes to the history of science and the history of agriculture, but perhaps more importantly to knowledge of how the one field interacted with the other. Scientific agriculture (or what might equally be described as agricultural science) is a subject that often falls through academic cracks, as Deborah Fitzgerald pointed out.[2] She noted that scholars are more likely to study either the history of science outside any agricultural context, or the history of agriculture and rural affairs with little reference to science generally. Jonathan Harwood recently suggested that the on-going poverty of science/agriculture material results from the fact that the subject requires knowledge of the history of science and technology combined, and work that spans science and technology is not particularly common.[3] The existing literature that does devote itself to science and agriculture in either North America or Europe does not tend to focus on how biology (genetics specifically) has affected farming.

Chemistry in relation to agriculture has attracted interest. The interface of American farmers and scientists with respect to agronomy and agricultural chemistry, for example, has received scholarly attention in relation to the spread of Justus Liebig's ideas in the United States between 1840 and 1880.[4] Concern with soil exhaustion triggered this nineteenth-century interest in chemistry and made agricultural experts urge farmers to keep livestock in order to fertilize the fields with manure. It was a common theme throughout the Anglo and European world. Agricultural experts, but not necessarily practical farmers, saw livestock husbandry as a device to produce more wheat or corn.[5] Much has been written about agricultural education and the role of science in it. Some of this material, namely, histories about the rise of experiment stations and agricultural colleges, deals with the way nineteenth-century, so-called experts attempted to convert North American farmers from their practical, but considerably less 'scientific' ways, to what farmers called 'book farming.'[6] Historians have assessed the characteristics of agricultural colleges in Germany and shown, for example, that research undertaken in German universities became less closely linked with actual farming practices over the early twentieth century.[7] Another study provides a comparative look at German and American orientation to genetic research for agricultural purposes.[8] This work does not, however, elucidate how that research and resulting education influenced breeding activities on farms. A recent review of agricultural research in France in the early twentieth century indicates its transformation via the introduction of the American structure of experiment stations and agricultural colleges. However, the study focuses more on the French concern with developing education in American biology and attitudes to genetics than on the application of such education to farming.[9]

Most of the material that does exist on the topic of genetics and its effects on farming, in either Europe or North America, addresses plant breeding.[10] The phenomenal success of hybrid corn in the marketplace has made the subject of its breeding particularly interesting. Historians have explained how genetics had 'scientized' the breeding of hybrid corn in the United States by the 1930s.[11] Hybrid corn and its use have interested scholars on other levels. The way hybrid corn-growing spread among farmers in the late 1930s, for example, triggered seminal sociological studies in the early 1940s, out of Iowa State University, on how innovation diffusion evolved. Did it result from the work of agricultural colleges through their education programs, from the spread of information by word of mouth, or from reading the farm press?[12] A number of excellent articles on plant breeding and Mendelism in the United States, France, and Germany appeared in the *Journal of the History of Biology*'s special 2006 issue.[13]

Genetics and its effects on animal breeding have received less attention than the plant-breeding situation. Some good work in the field does exist, though. A study of Mendelist research at an American experiment station, for example, shows how important chickens were to early genetic experiments.[14] European historians especially have approached the subject of livestock breeding in relation to genetics, and have done so increasingly in the past few years. Recent articles look at dairy-cattle breeding in the Netherlands and the effects of quantitative genetics on it over the twentieth century.[15] Another assesses early-nineteenth-century European livestock breeding before the advent of Mendel.[16] An excellent book looks at research on sheep breeding in Moravia from 1770 to 1870 and makes it clear how complicated the interrelationships were between Robert Bakewell's farm-animal-breeding principles, Darwinism, and the work of Gregor Mendel.[17] Scholars have looked at technology and its impact on animal breeding in the twentieth century from a biomedical point of view.[18] In spite of the existence of such excellent sources, it is still fair to say that, compared to other areas of history – including the history of science – there is a dearth of information on how genetics has influenced agricultural breeding, particularly of animals. As Harwood stated in 2006, 'The historical relations between biology and agriculture have received remarkably little attention.'[19] (Chicken-breeding practices have attracted scholars for other reasons, namely, to learn more about the enforcement of trade secrets or 'patenting,' that is, with the intellectual-property protection of biology.[20] As this book shows, it is difficult to separate issues of patenting from the infusion of genetics into agricultural livestock breeding.)

This book also provides information on the development of agricultural genetics, a subject that has received remarkably little attention in its own right from specialized historians of genetics. A comprehensive history of the development of agricultural genetics is badly needed. The relationship of classical genetics to molecular genetics, for example, needs more extensive research within the framework of agricultural genetics. Classical genetics, in the form of population and quantitative genetics, underpinned scientific chicken breeding as it evolved in North America and then globally, and it should be noted, at least in passing, that it still does. Molecular genetics (initiated by the discovery of the DNA structure by Watson and Crick in 1953) has done little (as of early 2010) to change breeding systems to create better birds.[21] That situation is in the process of changing, and will probably do so rapidly over the next ten years. Molecular genetics, in the form of genomics, may well revolutionize how selection methods work with all livestock breeding. SNPs (single nucleotide polymorphisms), found across the whole genome and constructed randomly into high-density groupings, can be used to assess the breeding potential of an individual by its DNA.[22] The technology is being used as a breeding selection tool for cattle. The Bovine Illumina SNP 50 Beadchip marker, commercially available by early 2009, helps locate the best dairy bulls.[23] By 2010 the test was being applied as well to beef bulls. The complex relationship between classical and molecular genetics is critical to the way agricultural genetics functions today, and the historical background to that development needs elucidation.

In the bibliography, I list the scientific literature on the subject of poultry breeding and genetics as primary material, but it could be argued that at least some of it served as secondary sources as well. Various governments in Canada published the reports on breeding made by experts in Britain and the United States. Sources often duplicated each other across countries. A study at the Maine experiment station would be noted in reports of the Ontario Agricultural College. An Alberta circular gave one of the clearest explanations of the first great American poultry breeder's mating systems – perhaps, in fact, clearer than the books this man wrote and published in the United States.[24] The Canadian fancy-poultry press provides the best detailed information on the British standards set for chickens, which the American Poultry Association would modify for North American use in the 1870s. No wonder I found myself reading American articles in Canadian publications, and Canadian articles in American sources, when it came to contemporary chicken affairs, scientific or otherwise.

The tone of the North American farm and poultry press changed considerably over the twentieth century. From the late nineteenth century until roughly the 1930s, articles described contemporary events, discussions that took place at pertinent meetings, and the breeding affairs of individuals. Such an approach gives immediacy to the past for the reader today. Fairly comprehensive reports on breeding theory arising from developing genetics became increasingly common after 1910. This became less evident after 1930. It is difficult, in fact, to study many agricultural trends in the second half of the twentieth century because of the poverty of first-hand information. As one archivist said to me, much of twentieth-century agriculture will be lost to us because of lack of basic documentation on many important issues. I attempted to overcome the problem by speaking to people who knew and know the industry from about 1940 until today, namely, breeders, geneticists, and government officials.

Chickens have fascinated people since time immemorial. The varied beauty of their feathering, the courage of cocks, the sweetness of fluffy chicks, the mothering of busy hens, and the wonderful food in the form of eggs and meat that they provide are all features that have endeared the species to us for thousands of years. Today we use chickens in a more mechanistic way than all other farm livestock. We follow practices that are not good for the birds and do not necessarily reflect well on us, in spite of the obvious benefits of such practices, namely, cheap and generally safe food for humanity's millions. Animal rights for the birds and questions around the methods we use to care for them, remuneration provided to the farmers who are egg or poultry-meat producers, and working conditions of people employed in processing plants are all issues that arise from the modern chicken industry and are of considerable importance, socially speaking. Such issues can be the venue for many studies of interest. This book draws attention to one of them. It uses chicken breeding in North America to look into the evolution of certain patterns in human thinking and developing knowledge; namely, how scientific and craft understanding of heredity laws interfaced with each other.

Chapter One

Historical Background: Chickens, Those Lowly Creatures

Chicken breeding can be described as the foundation or the 'basement' of the North American poultry industry, an agricultural livestock industry that has exhibited mushrooming growth over the twentieth century. Previously underpinned by poor marketing and processing structures, the poultry industry began a remarkable transition which gathered ever-increasing momentum between the late 1930s and the 1960s. Within that environment chicken breeding underwent what appeared to be a revolution. Geneticists took over control and introduced what was perceived to be new, 'scientifically' oriented approaches to the problem of breeding. The phenomenal success of these breeders and the concurrent apparent transition from farm know-how to 'scientific' breeding begs answers to certain questions. How did this happen? Was it genetics that altered chicken breeding so profoundly, first in North America and then globally? The questions are interesting, not only in their own right, but also because answers to them help illuminate a larger story: namely, the process of the scientific infiltration of agricultural production based on livestock. Before approaching these issues, though, we must see North American chicken breeding within the context of how its historical dynamics evolved before the birds had acquired even a semblance of importance, from either a craft/practical or scientific perspective. This chapter provides a general overview of what might be called an international 'pre-chicken-industry period,' that is, human relations with chickens and the growth of the poultry-breeding culture within the larger framework of European livestock breeding from the time of domestication until roughly 1850. Because British breeding attitudes and breeding cultural structures played a significant role in the foundations of North American livestock and poultry breeding, the situation in Britain is emphasized within the European framework.

Domestication and Early History

Until very recently it was generally believed that all modern chickens descend from one wild species, the red jungle fowl of South and East Asia named *Gallus gallus*. Genomics, however, suggests now that the domestic chicken represents some form of early hybrid breeding between the red jungle fowl and the grey jungle fowl, *Gallus sonneratii*.[1] As to where and when this took place, there is some disagreement. While it is possible that the Chinese raised birds from this genetic background over five thousand years ago, scholars are more likely to support the theory that true domestication of the species took place only about four thousand years ago (2000 BC) in the Indus valley. Early keepers of chickens used them for sport cock fighting, not food. The Aryans invaded the Indus valley about 1500 BC, quickly adopted the domesticated fowl for ceremonial purposes (although it might be noted that ceremonially used birds were normally eaten as well), and introduced the birds to other regions. The birds were taken to Persia, probably by the Aryans, and subsequently into Mesopotamia by the Persians. In the late sixth century BC the Medes and Persians brought poultry to Greece, where fowl continued to be valued for sport and ceremonial purposes. Only the poor consumed chicken meat. Through Greek and Persian influence, chickens spread both south (to North Africa) and west (to southern Italy, through Roman importation from their Greek colonies). Breeding for food, and not for sport pleasure or ceremonial use, appears to have been practised first by the Romans in Italy about two thousand years ago. These people quickly recognized that chickens could supply armies with easily transportable supplies. By the first century BC poultry keeping received considerable attention from Romans, and a number of treatises which dealt with sophisticated breeding concepts existed by then. The Romans created a heavy, meat-specialized bird and a lighter, egg-laying specialist as well. While they adopted chickens for food reasons almost as soon as the birds were encountered, the Romans used them for sport/ceremonial purposes as well.[2]

The way chickens entered northern Europe is not well understood. Some believe there were two avenues: via Iran and the Mediterranean. It is certain that the Romans played a role in expanding domestic avian territory and an increased interest in the birds, but domestic fowl existed in northern Europe in pre-Roman times. When Julius Caesar arrived in Britain in 55 BC, for example, he found that local people kept birds, generally game cocks of Asiatic background, and that cock fights were popular. He introduced the heavier-type chicken to Britain, but the Romans failed to make the birds important in a widespread way for food purposes in that country.[3] The eating of chicken meat was forbidden in Britain under Druidical law.[4] The birds continued to be bred for the sport of fighting throughout the Roman occupation. The fall of Rome spelt the death knell for any developing interest in chicken keeping and breeding for food purposes in Britain (and throughout Europe), and that fact probably at least partially explains why the decline of Druidical law and the rise of Christianity in Britain did little to reverse a lack of interest in chickens as suppliers of food.[5] Chickens acquired a lowly status that remained in place until the nineteenth century, even though long before that time chicken products were being consumed on a large scale. Perhaps it is for this reason that historical literature on poultry keeping and breeding for utility or agricultural, not sport, use in any area of Europe between Roman times and the nineteenth century is so limited. The collapse of the Roman Empire apparently had a devastating and enduring effect on the position of chickens within the perceived hierarchy of domestic food animals.[6] It has been argued that as early as 100 B.C., therefore even before the fall of Rome, chicken importance had begun to decline, and a sort of poultry 'dark ages' had set in.[7]

By the thirteenth century a complicated marketing system for eggs and poultry meat had evolved in Britain.[8] The birds were chiefly valued for their meat, and not regarded as first-rate egg layers.[9] Until well into the nineteenth century, however, British farmers paid little attention to the breeding of these food-producing birds. Fowl raised for food products were haphazardly bred, even though people who kept them did so for commercial, not self-sufficient, reasons. Poultry products, particularly meat, were considered delicacies, and for that reason production was rarely aimed at home consumption.[10] Bailey and Culley, for example, noted in 1805 that the birds 'were always articles *purposely bred to pamper luxury.*'[11] Any interest in poultry breeding that existed focused on the production of good fighting birds.[12] Even so, the literature on sport breeding before the eighteenth century is limited. Gervasse Markham wrote a treatise on fighting cocks, called *The Fighting-Cocke*, in 1615. His advice on how to select for breeding was governed by assessment of appearance. 'In your generall election, chuse him which is of strong shape, good colour, true valour, and of a most sharpe and ready heele,' he stated.[13]

1 The two original breeds of chickens kept in post-Roman Britain. The Fighting Game Cock to the left, and the Dorking to the right. Note the thick, meaty quality of the Dorking. This bird was used primarily for meat, not egg production. (*An Old English Game Cock Golden Duckwing*, by H. Atkinson, c. 1920, oil on canvas. Collection of W.C. Stevens, Melbourne, Australia. *Miss Fairhurst's Pair of White Dorkings*, by Ludlow, c. 1890. O'Shea Gallery)

Chickens in the Americas

It seems likely that chickens were not present in the Americas in pre-Columbian times. The Spanish introduced them to South America as early as the conquest period. The birds spread north to Central America and the southern part of North America from Spanish and Portuguese imports.[14] Domestic poultry were introduced to British holdings in the New World as early as 1609.

While the breeding of birds for commercial purposes commanded little interest in British settlers of the New World (as was the case in Britain itself at the time), the poultry-market situation was somewhat different in British-controlled North America. The settlers kept the birds for self-sufficient, not commercial, reasons. North American native peoples in the area quickly adopted domestic fowl as food too. The Iroquois of central New York State were known to have kept chickens as early as 1687. All people in the British colonies consumed poultry meat and eggs from domestic stock in the summer time (they utilized wild avian species as well), but poultry keepers did not see their agricultural flocks as vehicles to provide marketable commodities. The main commercial product that these birds yielded seemed to be feathers for beds and pillows. There was virtually no commercial market for meat or eggs from poultry in the United States before 1825.[15] No particular efforts were made, therefore, to improve the breeding of chickens for these commodities. Cock fighting was popular in the United States in a fairly widespread way until at least 1900, and good fighting birds always found a ready market for sport purposes, a trend as old as domestication itself. The unforgettable stories of 'Chicken George' in Alex Haley's *Roots*, published in 1976, bring to life how important cock fighting was to many in the United States between 1820 and the late 1870s. It is interesting to note, too, that although acts outlawing the sport were passed in various states over the nineteenth and twentieth centuries, it was not until August of 2008 that cock fighting became illegal by state law in Louisiana.

British Livestock Breeding Methodology:
The Work of Robert Bakewell

While the chicken commanded little attention with respect to breeding or care, British farmers wanted to know how they could improve their more important farm animals – by breeding and by care. The subject of care and husbandry took precedence over that of breeding, however, in written documents that have survived for the period before the eighteenth century. A few treatises dating back as early as Roman times and even earlier exist, but these too focused more on husbandry practices than breeding methodology.

The fundamental theoretical approaches, which would direct how nineteenth- and twentieth-century breeding of all livestock in the United States and Canada proceeded, evolved in the eighteenth century, that is, at the same time that empirical interest in natural history and heredity arose during the Enlightenment. Systematic descriptions of breeding methods which originated in Britain subsequently spread throughout northern and central Europe. These newly articulated ideas on the nature of heredity, and therefore on how to breed livestock, would fuse with long-standing practices followed by British cockfighting breeders to create empirical approaches to the breeding of all animals. These ideas ultimately shaped livestock and chicken-breeding strategies that developed in what had been and also continued to be British colonies in North America.

All aspects of farming changed in Britain over the eighteenth century, as people tried to make agriculture provide more food more efficiently. Within that framework, a breeding revolution with respect to livestock (if not chickens) occurred. A number of outstanding livestock breeders contributed to this phenomenon. The best known of these improving eighteenth-century breeders was an English tenant farmer, Robert Bakewell of Dishley.[16] In spite of wide recognition, Bakewell in fact left little written material concerning the methods he used or on his own systemized theory. Many since that time have believed that he was deliberately secretive. 'The mystery with which [Bakewell] is well known to have carried on every part of his business, and the various means which he employed to mislead the public' tended, John Sebright wrote, to undermine the importance of his work.[17] A small collection of letters between George Culley and Bakewell between 1787 and 1792, however, reveal something of Bakewell's thinking. Another contemporary, Arthur Young, also left information on the Bakewellian system for livestock breeding.

Bakewell attempted to improve the Longhorn cow and Leicester sheep for meat, and the size of the Shire horse. He succeeded in the case of the Leicester sheep, effectively creating that breed as early as the 1760s. His significance to the breeding world and to an empirical understanding of heredity, however, did not lie in breeds he developed or worked with. It lay instead in the breeding methodology he proposed and made famous. Bakewell might have practised breeding strategies that others besides him had worked out, but it is he who gets the credit for bringing their breeding knowledge to the public in a widespread way. The accumulative work of the British eighteenth-century breeders came to be seen, and defined, as the Bakewellian method. Results of breeding via this method had taught the breeders not only how to bring out the features they wanted to perpetuate in livestock, but also how to fix those features as well. The method provided a way to create 'breeds,' and it reflected a 'practical' understanding of the way heredity works.

Bakewell advised careful selection of males and females. He recognized the fact that both the male and female contributed to the make-up of their progeny at a time when only the best breeders did not reject female influence. He told the Elector of Saxony, as Bakewell reported to Culley in 1787, that 'the whole of the Art I was acquainted with was in [choosing] the best Males and best Females and keeping them in a thriving state.'[18] Good males bred to poor females would not, Bakewell knew, yield good results. The words 'keeping them in a thriving state' are important for an understanding of the Bakewellian method. The stock was to convert feed easily, but also must be vigorous.

One of the most important aspects of the Bakewellian method was an emphasis on sanguineous mating (that is, the mating of animals that were related) and the avoidance of any outcrosses. Arthur Young described the 'in-and-in' breeding theory embedded in the Bakewellian method.

> In breeding his bulls and cows (and it is the same with his sheep) he entirely set at naught the old idea of the necessity of variation from crosses; on the contrary, the sons cover the dams, and the sires their daughters, and their progeny equally good with no attention whatever to vary the race. The old systems in this respect he thinks erroneous and founded on opinion only, without attending either to reason or experience; and he asks anyone to point out a stock of beasts or sheep, now in high credit, that have not originated from this stock and bred in this way. Is it not the practice with the pigeon-fanciers and cockbreeders? And probably would hold equally good in horses and dogs, were they to preserve in the same fashion: but when anything contrary to their expectations occurs, it is attributed to the want of a proper cross in blood, and they immediately decline this mode. But ask whether anything has ever happened in this way that

they have not experienced in the common way of breeding; then why condemn the practice till further proof has been made by reversing every one of these maxims?[19]

These were innovative ideas for the times. Breeders in Britain and continental Europe of larger livestock had traditionally avoided sanguineous matings, often described simply as inbreeding by them, because experience had taught them that such practices could result in degeneration and lowered fertility. Producers of cattle, sheep, or horses realized, in other words, that increased relatedness, which brought about a move to increased uniformity in the hereditary make-up of individuals, might undermine vigour, and with this, fertility. It was a situation they wanted to avoid. But Bakewell argued that only through the mating of related stock for hereditary uniformity could animals be made to breed truly for improvement over time, and that careful selection at each generation could counteract negative effects of sanguineous mating. In the end, he made the principle of sanguineous union acceptable to thoughtful breeders trying to fix type in large livestock.[20] Breeders in Britain, who established early 'breeds' of livestock, particularly cattle, began to adopt the Bakewellian method. Charles and Robert Colling created the beef Shorthorn that way between 1780 and 1810, and Benjamin Tomkins the younger, who initiated such a program after 1790, produced Herefords.[21] Resistance to the practice continued, however, whenever people failed to understand the basic principle that sanguineous breeding could be done in different degrees, or when breeders did not recognize how critical the process of ongoing selection, generation after generation, was under this system. The secret to successful sanguineous breeding was careful selection.

Bakewell also advised another important strategy in the selection process. Stock would be judged on the basis of their progeny, namely, by what is called the progeny test. In order to judge the breeding ability of his males more comprehensively, Bakewell lent out bulls, rams, and stallions to neighbouring farms. He could test the progeny of these males under various conditions, as well as accumulate greater numbers on their productivity than would have been possible on his own farm alone. This situation gave him better predictability for future breeding. Males whose progeny he found acceptable returned to breed his chosen females, who, over time would be related to such males. In essence, Bakewell selected on the basis of ongoing vigour in order to maintain his blood-related lines, designed by the progeny test to intensify the traits that he desired. The special capacity of a given male mated to a given female to produce outstanding offspring enhanced the input of quality males and females.

While it is almost certain that Bakewell and his fellow improving, eighteenth-century breeders were aware of what livestock people and geneticists today call hybrid vigour (namely, the production of superior offspring by crossing two lines or breeds not related to each other), he did not breed for that feature. He wanted to establish stock that bred truly, and animals demonstrating strong hybrid vigour from outcross mating did not breed on to their type. Because Bakewell's animals had been selected by breeding to repeat good qualities beyond themselves, they stamped type on herds and flocks. Therefore, the use of Bakewell-bred males alone could improve a farmer's stock. Bakewell intended that to be the case and he hoped to force the farmer back to him for males to use for that purpose. His method of protecting his breeding work was simple. He tried to restrict the sale of females and to limit that of male animals for breeding purposes outside a circle of men working with him, namely, members of the Dishley Society. He focused particularly on the letting or renting out of rams. In 1791 Bakewell explained the situation to Culley. The 'Ram Company,' or Dishley Society, planned to pass a resolution by which Bakewell 'hop[ed] it [would] be agreed on not to let a Ram to any person (live where he will) but will engage not to sell any Rams but what he [should] see killed before they [went] out of his hands, or take any to market but what [was] disposed of for the season with such other regulations as shall be thought proper.'[22]

It is this sort of activity that made men like Sebright view Bakewell and his colleagues as secretive. The proposal to make males hard to come by and to restrict the movement of females struck many as simply the establishment of a marketing cartel. Farmers resented this blatant effort to skew the market in this fashion, and thereby control how improved breeding would be available to the public. In the spring of 1792, Culley reported resistance to the society's control over animals in this manner. As he explained: A group of farmers held a meeting and at it 'resolved unanimously, that a monopoly of any Trade, or Association, entered into by a set of Men (especially of those whose Resolutions enjoin secrecy) [were] highly injurious to the Public; and that Members of this Meeting [would] pursue every measure to counteract such Associations.' Farmers decided that the way to fight the monopoly was to avoid the stock of the Dishley Society and to do business only with other breeders.[23] Bakewell died in 1795, and the efforts to control Dishley blood in sheep lines petered out. By that time, however, Bakewell was largely credited with revolutionizing livestock breeding, in Britain and on the Continent, even if his attempts to affect the marketing of breeding by some form of patent had failed.

Livestock and Chicken Breeding: The Ideas of Sir John Sebright

The Bakewellian principles of inbreeding, as described by Arthur Young, clearly dovetailed with those of fighting-cock and fancy pigeon breeders in Britain. It is possible to see how closely Bakewell's method matched or differed from contemporary British chicken-breeding practices for game cocks by looking at the words of a late-eighteenth- and early-nineteenth-century chicken breeder, Sir John Sebright. Because his explanation of breeding principles in relation to the Bakewellian method came to provide the rationale behind chicken craft breeding in North America, it is worth looking in considerable detail at Sebright's breeding theories, which he described in a pamphlet, *The Art of Improving the Breeds of Domestic Animals*, published in 1809.

For Sebright, breeding by ancestry should be done in conjunction with breeding by the progeny test. Animals were more likely to perpetuate quality, Sebright argued, if generations behind them had already done so. He believed that animals would not breed well with each other unless they came from known quality ancestry, because quality ancestry suggested the ability to pass on quality. 'We should not breed from an animal, however excellent, unless we can ascertain it to be what is called *well bred*; that is descended from a race of ancestors, who have, through several generations, possessed, in high degree, the properties which it is our object to obtain,' he advised.[24] He also advised the selection of males and females in relation to how their strengths complemented the weaknesses of the other. Men were poor breeders, he explained, when they thought that by picking good males and females and mating them, they had done everything necessary to produce good offspring. 'This is not the case,' Sebright stated, and elaborated as follows: 'Were I to define what is called the art of breeding, I should say, that it consists in the selection of males and females, intended to breed together, in reference to each other's merits and defects. It is not always putting the best male to the best female, that the best produce will be obtained.'[25] Sebright knew that equal input came from male and female, and therefore that one could correct the defects of the other.

Inbreeding was at the heart of Sebright's, as well as Bakewell's, methodology of breeding for lines that perpetuated improvement truly generation after generation. The theory was that such breeding practices enhanced certain hereditary features and eliminated others, trends that could not be accomplished by breeding animals unrelated to each other. Sebright fully understood the fact that by crossing two breeds or strains, the progeny of the first cross might be good, but they would not breed truly.[26] While he recognized that sanguineous mating produced lines that bred true, it was equally clear to him that the bad could be strengthened as well as the good from inbreeding. 'By breeding *in-and-in*, [a] defect, however small it may be at first, will increase in every succeeding generation; and will, at last predominate to such a degree, as to render the breed of little value,' he said.[27]

It seemed to Sebright that resistance to Bakewellian principles arose out of misunderstandings over the meaning of certain expressions (particularly that of inbreeding or breeding 'in-and-in'), not over the value of mating animals that were related to each other. Sebright viewed breeding 'in-and-in' as the most intense form of inbreeding possible, and to him it meant only sister-to-brother matings. Bakewell's 'in-and-in' breeding clearly meant father-to-daughter or son-to-mother mating, and Sebright would not have labelled this method as breeding 'in-and-in,' or inbreeding. For him such unions should be described as line breeding, a method commonly used at the time by pigeon and poultry breeders. 'This is not what I consider as breeding *in-and-in*,' he explained, 'for the daughter is only half the blood of the father, and will probably partake, to a degree, of the properties of the mother.' Regardless of a definition for line breeding or its relationship to inbreeding or breeding in-and-in, Sebright held that 'Mr. Bakewell had certainly the merit of destroying the absurd prejudices which formerly prevailed against breeding from animals, between whom there was any degree of relationship; had this opinion been universally acted upon, no one could have been said to be possessed of a particular breed, good or bad; for the produce of one year would have been dissimilar to that of another, and we would have availed ourselves but little of an animal of superior merit, that we might have had the good fortune to possess.'[28]

Sebright's thoughts revealed a particularly significant conundrum that would be embedded in all breeding exercises – practical and genetic – in the future: namely, the meaning of inbreeding (or breeding 'in-and-in') and the separation of it from the meaning of line breeding. He (incorrectly) believed that father/daughter or mother/son combinations represented less severe forms of inbreeding than brother/sister matings. But by designating the former as line breeding he set an example that breeders would follow until the present day: namely, while inbreeding generally was considered dangerous, forms of inbreeding that became acceptable to breeders should be defined as line breeding. Because the meaning behind the term line breeding could be applied to various forms of inbreeding (and was over the years, in accordance with how breeders chose to define different levels of inbreeding), confusion was built into the definition. In 1910 the *American Poultry Journal*, for example, commented on the ongoing riddle, but only added to the confusion surrounding the problem.[29]

Sebright outlined an elaborate mating system that controlled levels of inbreeding (defined by him as line breeding) on a percentage basis. 'If the original male and female were of different families, by breeding from the mother and the son, and again from the male produce and the mother, and from the father and daughter in the same way. Two families sufficiently distinct might be obtained, for the son is only half the father's blood, and the produce from the mother and son will be six parts of the mother and two of the father.'[30] His theory of inbreeding as a quantitative and percentage issue would be important to craft chicken breeders later in the nineteenth century, and geneticists as well in the twentieth.

The Bakewellian method, which revolutionized large-livestock breeding, effectively fused the older poultry breeding practices of inbreeding with new selection criteria. One might say that poultry breeding strategies entered the world of large livestock production through the Bakewellian system. Those interested in fowl breeding at the time found it difficult to see what was innovative in Bakewellian principles for poultry breeders because inbreeding was not new to them, as Arthur Young's comments make clear. Young stated that the mating of father to daughter and mother to son was commonly practised by producers of fancy pigeons and game cocks. The early-eighteenth-century farm press in Britain gave advice to those who might take up an interest in the breeding of poultry, and suggested the same allegiance to inbreeding for poultry production, or breeding within families rather than outcrossing. In 1732 *The Country Gentleman and Farmer's Monthly Director* advised, 'Leave only a Cock to seven or eight Hens, that your Summer Breed may be strong; but if you have not sufficient number of Cocks for the Hens, rather sell off some of the Hens than buy in Cocks which are Strangers, for there will be disagreement which will occasion a weakness and poorness in the chickens.'[31] 'Stranger' cocks seemed to imply unrelated males as much as unknown ones.

The Organization of Breeding, 1700–1900

Improving European breeders of large livestock over the late eighteenth and early nineteenth centuries established ways of organizing the results of their breeding in order to distinguish the animals from the run-of-the-mill beasts. The breeders described their stock as thoroughbred, purebred, or standardbred and argued that certain breeding practices had to be followed in order to warrant such a definition. In the beginning, the definitions implied the use of a particular selection methodology for breeding, but in the end all three came simply to label animals conforming to regulations set out by breeders.

The earliest theoretical breeding framework put in place for the improvement of animals was the Thoroughbred system, and in its pure and earliest form it applied only to the production of the Thoroughbred horse in Britain. Much of the culture and philosophy within the Thoroughbred system, however, went on to influence the workings of both standardbred and purebred breeding. The legacy of Thoroughbred theory, and in some cases even breeding approaches, has been profound. Much of its eighteenth-century attitudes became part of chicken/poultry culture of North America in the late nineteenth century, and 'thoroughbred' thinking still plays a role in many sectors of the present larger world of animal breeding.

In seventeenth- and early-eighteenth-century Britain, a more concerted effort to breed good riding animals developed in order to supply the cavalry with superior mounts. The breeding of war horses, or 'remounts,' had been a serious issue in continental Europe for centuries by this time, but sea-bound Britain had not felt the same pressing need for fast, capable cavalry mounts. The British nobility began to take more interest in horses for the country's defence after King James I imported an Arabian stallion for breeding purposes in 1616. More Arabian stallions came in over the seventeenth century, and when these were crossed on local mares, superior animals resulted. It appeared to the breeders that the Arabian stamped good type on stock with great consistency, and the imported stallions commanded deep respect. Their potency, it was believed, reflected the 'purity' of their background. Because the thought of the day (it was before Bakewell's time) in most livestock circles went against any form of sanguineous breeding and because the Arabian was rare compared to native horses, breeders began to keep records more systematically of their stock's genealogy. In that way they could prove that they had avoided sanguineous breeding and also that the animals resulted from an Arabian cross on a native mare. These half-Arabians could be recrossed on Arabians as well, as long as no inbreeding was practised. Stock that came out of such crosses were said to be 'thoroughly' bred, therefore 'thoroughbred.' The horses were tested for stamina by racing them over distances as long as eight miles. Breeding via the progeny test would, under these circumstances, become part of a breeder's strategy. Only near the end of the eighteenth century did sprinting for sport become more common. The move away from utility motives in racing worried a number of breeders. The Thoroughbred had been designed to be a working horse.

The increased emphasis on racing meant that regulations had to be put in place in order to stop fraudulent entries – for example, entering a four-year-old in a two-year-olds' race. In 1791 James Weatherby compiled information on racing horses and published the material in what was known as the General Stud Book (GSB), the first public registry system in the world. Weatherby noted sires and dams of horses if he was able to do so, but breeders continued to rely on their own records for breeding purposes. Public pedigrees identified stock, thereby making it more valuable for a purchaser, especially if the buyer lived some distance from the seller. Animals carrying a public pedigree also stood apart from the common equine crowd. Elitism became attached to the new Thoroughbred because of its identification process and its genealogical background. The animal was seen as 'pure,' 'thoroughly' bred, and stamped as a 'blood horse.' Inbreeding had been avoided, generally speaking, in the creation of the 'pure' horse. Public pedigrees, if not breeding tools, acted as guarantees of both purity and a new style.[32] The prevalence of limited inbreeding, however, ensured a rather widespread unevenness of type in these early 'Thoroughbreds.' Only after breeders abandoned the formula of the pure Arabian/local cross (in the late eighteenth century) and began to work with the pool of horses that had resulted from such matings, did the Thoroughbred begin to take on real uniformity of type. Restricting the hereditary background of the animals used for breeding in this fashion would have compelled at least limited inbreeding. What is most important about historical Thoroughbred horse breeding for the story of how all future animal breeding developed is the fact that certain characteristics could be attached to animals bred in a certain fashion. The evolution of the Thoroughbred horse introduced the idea of a public registry system and the theory of purity. The legacy of the Thoroughbred for all future breeding would be an allegiance to recording pedigrees and the belief that pedigrees labelled animals as pure. Neither, ultimately, related to actual breeding methods.

The eighteenth-century improving European breeders introduced the idea that inbreeding stamped type, and that the way to improve stock was to progeny test. These principles were at the heart of the Bakewellian method. Both principles differed profoundly from those behind Thoroughbred horse breeding, where any sanguineous mating was anathema and genealogy, as much or even more than the progeny test, served as the basis for selection. Something new entered the breeding world, however, when aspects of horse-breeding culture fused with the Bakewellian system within the world of British beef Shorthorn cattle.

Robert and Charles Colling had created the beef Shorthorn using the Bakewellian method between 1780 and 1810. Shifts in British agriculture over the eighteenth century had increased the interest of the land-owning nobles in better farming methods, and under these conditions the improved Shorthorn caught their attention. They began to buy Shorthorns. By 1830 the thinking behind Shorthorn breeding had changed from true Bakewellianism as a result of the work of one man in particular, Thomas Bates, who probably reacted to this support of the elite. Bates began to breed Shorthorns with an emphasis on genealogy (Thoroughbred horse methods) rather than the progeny test (Bakewellian methods). He continued inbreeding and inbred by mating brother to sister, but unlike Bakewell, he stated that he practised sanguineous mating to preserve purity. Bates linked sanguineous breeding with the Thoroughbred idea of purity. After George Coates set up a public registry for Shorthorns in 1822 (the world's second public registry system), Bates attached publicly registered pedigrees to both purity issues and breeding methods. He introduced a new feature to public recording. Public pedigrees, not breeder records, should be used as breeding tools. By the time Bates died in 1850, a new system, based on his innovations and Bakewellian thought combined with Thoroughbred horse culture, had evolved: namely, the purebred structure.

An animal was 'pure' to type, that is, purebred if it was acceptable by genealogy for recording in a public registry. Sanguineous breeding had initially been part of the guarantee of purity, and proof of that breeding could be seen in pedigrees. For some time over the nineteenth century, breeding outlook demanded that purebred registered stock be inbred. Bates, for example, marketed his Duchess line of Shorthorn cattle on the basis of purity because of intense inbreeding as recorded in pedigrees. No outcross had 'contaminated' them.[33] With the better fixing of type, however, any form of inbreeding became less significant. The purebred system in its mature form simply meant that an animal could carry a pedigree if both its parents had been recorded in the stud book. Pedigrees guaranteed consistency of type because of the restricted hereditary background that had been established by earlier sanguineous breeding. Any method of breeding was acceptable within that framework – inbreeding and outcrossing, selection by the progeny test or ancestry. The philosophy behind Thoroughbred horse breeding continued to shape attitudes to purebred stock, regardless of breeding methods. Purity was linked to genealogy, but not to any particular breeding method or selection process used to generate that genealogy.

While 'thoroughbred' thinking shaped the ethos of purebred thinking, purebred breeding, in turn, became infused with the breeding of the Thoroughbred horse. By the late eighteenth century, a horse could be labelled a Thoroughbred if both its parents were recorded as Thoroughbreds in the stud book, and not only if the animal resulted from a cross to an Arabian stallion. (Arabians continued to be registered in the GSB until the 1960s, a hangover from the early days when the Thoroughbred breed was developing.) Breeders came to rely on the GSB as a breeding regulation, adhering to purebred principles. The Thoroughbred had become a 'purebred' horse. Sanguineous breeding within that GSB-limiting horse population continued to be culturally unacceptable for some time, but it was up to the breeder if he wanted to use the method. Inbreeding did in fact become more common over the nineteenth century. The idea of 'pureness' in Thoroughbreds reached a new height in 1913, when Lord Jersey managed to convince horse breeders in Britain to use only animals registered in the British GSB for breeding purposes. For over thirty years, racehorses from the United States and France of outstanding quality were not eligible for GSB pedigrees (because their sires and dams did not have GSB papers) and therefore could not be used in Britain for breeding.[34] Man o' War, or Big Red, the famous American racehorse, and therefore his progeny, did not pass the Jersey test. Purity in this case related to marketing issues that had nothing to do with breeding methods.

European systems of stipulating 'breed' in relation to horses would be particularly important to the breeding outlook pertaining to poultry over the eighteenth century in Britain and North America. These differed from the British Thoroughbred and purebred structures. Methods that regulated the development of the European Warmblood horses reflected the idea that stock should meet a certain standard in order to qualify for a 'breed' label. Warmbloods resulted from the crossing of the so-called cold blood of any heavy draft horse with the so-called hot blood of an Arabian or Thoroughbred. Various Warmbloods were produced through this type of crossing – the Trakehner, the Hanoverian, and the Dutch Warmblood (which was subdivided into three styles: the Gelderlander, Groninger, and the Tuigpaard). Warmbloods received accreditation as a 'breed' on the basis of type assessed by inspection and testing for performance, not genealogy as recorded in pedigrees. The horses were bred to a standard. The methodology used in breeding mattered even less than their genetic background.[35] Even though Warmbloods were created under a standardbred structure, in the nineteenth century, the use of pedigrees for breeding purposes began to infiltrate the system. And with time, purebred breeding philosophy increasingly governed how Warmbloods could be pedigreed. While inspection and testing for performance continued, those animals seeking admission to the stud book increasingly came from the genealogical background of accepted, pedigreed horses. Performance testing, however, counteracted the effects of purity philosophy that so permeated the purebred and Thoroughbred systems. Warmbloods were (and are) viewed as capable horses, not horses of 'pure' breeding. Under these conditions, breeding methods used to create them remained as flexible as they had always been.

One of the best known animals created under a standardbred system was the Standardbred trotting/pacing horse. Different North American lines of trotting and pacing animals could be registered in a stud book as early as 1839, but recording was done on the basis of their speed. The horses were bred to a speed standard, and thus labelled as Standardbred. Speed standards were more formally organized and defined later in the nineteenth century. By the late 1870s any stallion, mare, or gelding (castrated stallion) that could trot or pace at the rate of a mile in two and a half minutes under set racing trials organized by the National Trotting Association could be entered in the Breeders' Trotting Stud Book. An emphasis on breeding to a standard appeared to some to be more effective than purebred or thoroughbred breeding. It was noted that the speed of Thoroughbreds increased only 8 per cent over a twenty-five-year period (1875–1908), while Standardbreds ran faster at the much higher rate of 27 per cent over a twenty-five-year period (1877–1913). In being bred to a standard the trotting/pacing horses had increased their capacity to run by half a minute over less than fifty years.[36]

Even so, the purebred breeding ethos tended to infiltrate many standardbred systems over the late nineteenth century, and did so with respect to this trotting/pacing horse. By the 1880s, entry into the book was possible on the basis of genealogy that indicated potential speed. Any horse sired by a registered Standardbred and from a mare sired by a registered Standardbred qualified for entry into the stud book. Any mare that was sired by a Standardbred stallion or herself had produced a tested Standardbred also qualified. By late 1898 entry into the stud book had become purely an issue of genealogy, not one of performance testing. Any animal whose sire and dam were registered as a Standardbred qualified for recording.[37] Standardbred breeders could select any method they chose in order to breed. The only restriction was that they had to breed from a registered stallion and a registered mare.

In the end, the fundamental theoretical stance of the thoroughbred, purebred, and standardbred systems became entangled with each other in North American breeding circles. Even the words 'thoroughbred,' 'purebred,' and 'standardbred' were erroneously used interchangeably to describe the same thing. In reality, all described purebred breeding, which held hegemony over the other two by the mid-nineteenth century. It was, however, still perceived to be an organizing, not a breeding system. That situation changed rapidly before the twentieth century began.

By the late nineteenth century, while the North American farm press might consistently tout purebred breeding as *the* method to breed by, actual methodology describing breeding in the press outlined either the Bakewellian method or the strategies of the Collings.[38] The 'science' of breeding taught in North American agricultural colleges in the late nineteenth century focused on how these eighteenth-century breeders practised the 'art' of breeding, but defined these methods to purebred principles.[39] By the late nineteenth century neither the press nor the agricultural experts seemed to recognize the fact that the purebred system had nothing to do with actual breeding and did not reflect either Bakewell's method or the breeding strategies of the Collings.

Chicken Breeding in Britain and North America by the Mid-Nineteenth Century

Over the same period that the purebred and standardbred systems were evolving, the chicken-breeding situation in both Britain and North America began to change. Ultimately, the outlook of both breeding systems would play a role in the way new approaches to chicken breeding over the late nineteenth and early twentieth centuries developed, but other factors were also part of the story. Cock fighting was outlawed in Britain in 1849, for example, and that factor shunted British interest in the breeding of chickens into other avenues. (It should be noted, however, that cock fighting in Britain did not end at that time. The sport was not outlawed in Scotland until 1895.) The growth of poultry shows in Britain – the first took place in London in 1844 – fuelled an interest in breeding and in breeding-for-beauty competition. An infusion of new varieties from Asia occurred at the same time, allowing for an expanded breeding base. About 1845 the Shanghai or Cochin was introduced, and the Indian Game (which when crossed with the old English Game became the Cornish) in 1858.[40] (Some poultry experts argued that the Indian Game was known to Englishmen from the time of Britain's contact with India.)[41] Changes in British life further encouraged the keeping of the birds – urbanization and industrialization moved many rural people into cities, and poultry could be raised under such confined conditions. Popular poultry shows, the introduction of new genetics, and interest in breeding arose in North America at the same time.[42] Much of North American breeding orientation and organization would be derived from the British scene, even though Britain and North America obtained the new varieties used for innovative breeding (from Asia, India, and the Mediterranean) independently.

When a new emphasis on chicken breeding began in the late 1840s, people who flocked to the cause of breeding better poultry tended to favour the organization structure of purebred breeding. George Burnham, an important early American poultry breeder and importer who played a significant role in the rising popularity of poultry exhibitions in both the United States and Britain, had the following to say about poultry and pedigrees:

> In England, amongst other nonsense bearing upon this subject [of hen fever-struck and ignorant people], the more cunning poultry-keepers resorted to furnishing the pedigrees for the birds they sold. This trick worked with admiration in Great Britain for a time, and the highest-sounding names were given to certain fowls, the progeny of which ('with pedigree attached') commanded the most extravagant and ruinous prices, in the English 'fancy' market ...

Burnham found that his American customers often demanded pedigrees. He quoted a certain letter addressed to him on the subject:

2 A modern White Cochin. This was a popular meat and exhibition bird late in the nineteenth century. It is a very large breed of chicken, and therefore provided much meat. Its feathery legs attracted those interested in beauty. (Photo by Pete Paterson)

'I have been a live-stock breeder for some years in this and the old country, and I was desirous to obtain only *pure*-blooded fowls when I ordered the "Cochins" of you last month. I asked you for their *pedigree*. You have sent none ... Am I dealing with a gentleman? Or are you a mere shambles-huckster? What are those fowls bred from? ... Who *are* you? I sent for a pedigree and I want one. *I must have it*, sir. You will comprehend, I assume. If you do not, I can enlighten you further. In haste, —.'[43]

Burnham smiled at the letter – especially because the writer was known for supplying elaborate pedigrees when he sold stock. Burnham answered by saying he would supply his own pedigree, which started with a man named Adam.[44] The idea of 'purity' in fowl thoroughly amused Burnham. He relayed a conversation he had had with a 'notorious shark' on the subject.

'There is one thing you should always bear in mind,' said a notorious shark to me, one day, while we conversed upon the subject of breeding livestock successfully – 'there is one thing you should always remember; and that is, under no circumstances ever permit a fowl or a pig to pass out of your hands to a purchaser, unless you *know* him to be of *pure blood*.' This is a pretty theory, and I have no doubt, such a course would work to admiration, if faithfully carried out (as *I* always intended to do, by the way); but in this country this was easier to talk about than accomplish.[45]

Burnham might have been amused by the reiterated demands of potential buyers on purity of lineage, but it appears he was not above deceiving people about the background of his stock, a problem considered to be serious among poultry breeders. The British fancier and Brahma breeder, L. Wright, claimed that Burnham had deliberately bred Cochins to look like true Brahmas, and that his claim to be the first Brahma breeder was false. 'It appears that ... Mr. Burnham had no real Brahmas in his possession, but having a large number of Cochins, of all colours, that he endeavoured to *imitate* for business purposes a fowl he found so popular and valuable, at the same time being perfectly aware of the great difference between the real strain and his own,' wrote Wright in 1870.[46] It was important that a bird called a Brahma was indeed a Brahma and not one simply bred from other breeds to look like one. Breeders normally attempted to protect their breeding through organizations in order to distinguish their work from those who made false claims about the breeding background of stock. Fraudulent labelling was one reason that breeders organized themselves into associations. Formalized structures designed to support the breeding of chickens and protect the breeders evolved rapidly after the mid-nineteenth century in Britain and North America, as will be evident in the next chapter. Chickens were also about to undergo a metamorphosis which led to the shedding of their lowly status. All chickens – like horses, cattle, sheep, and hogs – were to be improved, and breeding organizations were designed to encourage the spread of better breeding methods that practical breeders developed on the basis of trial and error, that is, through experience.

Chapter Two

From Barnyard Scavenger on North American Farms to Bird of Beauty and Use

With a rising emphasis after 1850 on establishing better chickens, breeders in North America adopted the methods of the eighteenth-century fighting cock and fancy pigeon breeders. Crosses of 'breeds' and 'types,' combined with blood-related matings, laid the groundwork for the development of the most important varieties for both show and utility stock over the late nineteenth and early twentieth centuries, as farmers tried to find a useful bird, and fanciers tried to fix style.[1] A new focus on chicken breeding changed the status of the birds. Increasingly viewed less as barnyard scavengers after 1850, chickens encouraged the rise of organizational ways of orchestrating breeding, the establishment of various breeding systems, and a societal structure within which breeding took place. This chapter explores these developments, which explain the nature of craft-breeding methodology, theory, and culture as they evolved in the United States and Canada. It will then be clear how complicated and sophisticated craft breeding and culture were before genetics catapulted into this livestock world.

Chicken Breeds

Individual chicken breeds would dominate, first, craft breeding in North America and, subsequently, genetic breeding around the world. The historical background to their development is, therefore, part of the craft/genetic-breeding story. The Barred Plymouth Rock, the White Plymouth Rock (which was a mutant of the barred variety), and the American White Leghorn, all of which evolved in the United States through the breeding methods applied by early craft breeders, became the most popular breeds of the pre-genetic era and later the most significant for global chicken production under genetic breeding. It would be fundamentally these breeds (along with the input of several others developed later, most particularly the Cornish and the Rhode Island Red) that geneticists used.

American breeders created Barred Plymouth Rocks by crossing a number of different types of birds, then inbreeding to fix type. Foreign blood would be introduced from time to time to strengthen certain characteristics weakened by inbreeding. The result was a useful bird, even if early lines did not always produce consistent results as to type. Dual-purpose (meaning good for both egg and meat production), the Barred Rocks incubated their eggs, and were therefore known as sitters. Vigorous, active, and with a foraging instinct, the Rocks were also large birds, relatively speaking, with some muscle development into the legs. Enhanced muscling made them good meat birds. Because they provided both meat and eggs, incubated their eggs, and were able to fend for themselves naturally, they quickly became favourites of farmers in the United States and Canada.

The first bird to be called a Barred Plymouth Rock seemed to result from the crossing of the Cochin, Malay, and Dorking, and was labelled 'a mongrel of little worth.' Dr J.C. Bennett claimed to be the creator of this stock in 1849, saying that his birds had one-half Cochin, one-quarter Dorking, one-eighth Malay, and one-eighth Indian Game blood. Other men stated that they had been Plymouth Rock originators. The 'Battle of the Rocks' controversy raged in New England between 1872 and 1875. It would appear that a number of centres existed in Massachusetts and Connecticut where 'Plymouth Rocks' were bred.[2] These various types all disappeared. The foundation strains of the Barred Plymouth Rock bird that endured arose between 1864 and 1874, from an original Java hen/Dominique cock cross, recrossed on a White Brahma. D.A. Upham first exhibited the new type at shows in 1868–9. A more complicated crossing system went into the final make-up of the Barred Plymouth Rock, fixed in type better by 1878. Early Rocks, however, aroused the scorn of many people. Comments such as 'Plymouth Rocks are not a breed and never will be' or 'They are a miserable mongrel' were common.[3] Crossing and recrossing the developed original strain on dark Brahmas, black Java, and some Minorca stabilized and improved the Barred Rocks, although some continued to see the bird as the American mongrel. The breed entered Britain as early as 1879. The white variety arose from the hatching eggs of the barred variety in 1875, when O.F. Frost of Monmouth, Maine, found he had white chicks from his barred eggs.[4] White Rocks had all the good qualities of the Barred Rocks.[5]

One of U. R. Fishel's White Plymouth Rock male birds which he intends showing at the leading shows the coming winter. There is no doubt but what the U. R. Fishel White Plymouth Rocks are the most popular and profitable breed of fowls we have today; not only has Mr. Fishel bred his White Plymouth Rocks to a high state of perfection in fancy requirements, but as a commercial fowl, as an egg producer and as a table fowl there is no breed of fowls to compare with them. Two eggs a day from a Fishel White Plymouth Rock hen has been reported quite often the past season. For further information regarding this wonderful strain of fowls we beg you to note Mr. Fishel's advertisement in this issue.

3 An exhibition White Plymouth Rock, bred by U.R. Fishel, 1908. The White Plymouth Rock is one of the foundation breeds of the modern broiler industry. The breed was a mutant variety of the Barred Plymouth Rock. Note the thick, compact body form, which suggests good meat quality. Meat quality can be seen down into the legs, adding to the value. Fancy feathering is absent, when compared to the Cochin. (*American Poultry Journal*, August 1908, 585)

The Leghorn, designed primarily for egg laying and therefore as a single-purpose bird, originated in Italy, where for generations it had been bred for egg laying. Leghorn types might have existed as early as Roman times. Introduced to North America before Britain, the white variety arrived first, in 1828.[6] Vigorous and active birds, Leghorns did not (and do not) incubate their eggs, and are called, as a result, non-sitters. This was not particularly disadvantageous in warm climates like Italy's, where incubation was possible without sitting. While needing some form of artificial incubation in a colder environment, their active nature helped them flourish in colder climates when they left their native Italy, because movement kept them warm.[7] In 1835 a poultryman from New York City imported a few Brown Leghorns. White Leghorns have often been thought to have originated as mutants of the brown variety in Italy. White Leghorns, however, could be found in Italy as easily as the brown. In North America the white variety soon dominated. The crossing of the white Minorca on Leghorns, and the importation of more white Leghorns between 1840 and 1845 ensured that trend. Breeding strategies in the United States led to an even more productive egg-laying bird. By the 1870s American White Leghorns had arrived in Britain.[8] The Leghorn had drawbacks, though, when compared to the Barred Rock. It was a non-sitter, a small bird that did not readily fatten, and therefore a poor provider of meat. Its advantages over the Barred Rock were its capacity for superior egg-laying, and its greater ability to breed truly for differing strains within the one breed. Since the Leghorn was well established as a breed before it entered North America, it also conformed readily to type more easily than the Barred Rock.

Early Organization of Chicken Breeding

Formal organization of chicken breeding began in Britain before it did in North America, and developments in Britain would have a profound impact on the poultry breeding situation in the New World. Artisans in the counties of Lancashire and Yorkshire, who were primarily concerned with creating beautiful birds for recreational (not agricultural) reasons, organized breeding around a show system. The sport of exhibiting birds expanded rapidly in Britain, and soon required a more formal structure to regulate it. In 1865 and 1866 the London Poultry Club created a document called the Standard of Excellence, which set a way to measure the quality of individuals within each breed. This situation made it easier at exhibitions to judge fowl. The system worked off fifteen points devoted to shape, feathers, overall balance, and tone of colour. No rules were set for breeding methodology. Cross-breeding was an acceptable way to establish type. The acid test did not revolve around the genealogical purity of a bird's background, but rather around its ability to match the standard. The era of the 'standardbred' fowl had begun.

A.M. Halsted of New York printed the British document in 1867, reprinted it in 1870, and revised it in 1871 to work off one hundred points rather than fifteen. That year William H. Lockwood, of the Connecticut State Poultry Society, compiled a new set of standards, labelled the American Standard of Excellence (which also worked off the 100-point system) owing to the way Halsted had dealt with the Barred Plymouth Rock. Dissatisfaction with these early standards led in 1873 to the formation of the American Poultry Association in Buffalo, New York, by poultrymen from both the United States and Canada. The association produced a new workable Standard of Excellence in 1874 for North America, also based on one hundred points, to be used for exhibition purposes and judging. A larger edition appeared in 1875, and in 1888 the name was changed to the Standard of Perfection. The document tended to go for revision every eight years.[9] It is important to note that the American Poultry Association was a North American, not simply an American organization, and also that it was established by the breeders themselves. The American Poultry Association organized chicken breeding in both Canada and the United States.

Chicken breeding might have developed from the idea of breeding to a standard, or standardbred breeding, but concern over introducing foreign blood became widespread in poultry circles quite early, indicating that an allegiance to a purebred ethos always underlay that of standardbred breeding. Important reasoning supported the incorporation of purebred breeding philosophy into the standardbred system. Since the standards were designed to dictate what results to breed for and did not stipulate what genealogy went into breeding, it was evident almost immediately that theoretically any material could be used to make a new set of birds meet that standard. Under these conditions there would be virtually no way for breeders to protect the value of their efforts. Quite quickly everyone agreed that one should not be permitted to re-manufacture established type from a new and different genetic base, and that such activity should be viewed as fraudulent. (That explains why Burnham's Brahmas were not seen as either authentic or pure even if up to the standards.) Purebred breeding relied on proven genealogical background, thereby giving evidence of what stock had been used in breeding programs. It could, therefore, act as a

4 An advertisement for the 1906 version of the American Standard of Perfection. (*American Poultry Journal*, July 1907, 635)

patent on the work of breeders, and, accordingly, breeders of standard-bred chickens often described their stock as purebred. Basic to purebred breeding, however, is the establishment of a pedigree recording system, something that chicken breeders did not have. Some began to think there should be one.

As early as 1870 the *Canadian Poultry Chronicle* openly campaigned for a public registry system for poultry in order to make them truly qualify as purebred. 'Poultry with a registered pedigree would be a novelty, yet we see no reason why the novelty should not exist; other livestock have their Pedigree-Book, and why not poultry too?' the paper asked. Pedigrees would solve another problem too, the journal argued, namely, what the origin of evolving breeds actually was. 'Is the origin [of many American breeds] to remain buried in obscurity [as is true in Britain] as it now apparently is? Surely not,' remarked the *Chronicle*.[10] In 1874 the *Poultry World* in the United States tried to get support for a registry which would pedigree standardbred poultry, but ridicule from other poultry journals killed the idea.[11] Purebred-breeding sentiment in poultry circles continued after that time and often underlay standardbred-breeding sentiment. Comments to the effect that ancestors count, and pedigrees matter because they keep track of ancestors and also command higher prices appeared in the poultry press.[12] The slow but relentless pressure to move towards 'purebred' poultry can be seen in ongoing campaigns against the mongrel or 'scrub' hen. Publicly funded programs to remove these birds from farm flocks became common.[13] Never use scrub over 'purebred' poultry, stated a bulletin released by the Alberta government. 'To breed pure is to mate birds of the same breed, and to mate crossbreds means the production of scrubs, and to attempt successful poultry raising with scrubs results in decided failure,' the bulletin explained.[14] Cultural considerations had come to overlay Sebright's empirical thinking when it came to the reasoning behind breeding lines of known background. Sebright wanted to perpetuate quality, and saw attention to ancestry as a method of restricting the inflow of unknown hereditary characteristics. His interest in ancestry did not arise from concern with emotionally charged ideas that seemed to assign birds to social classes.

Breeding for Beauty and/or Use: The Role of Exhibitions

The organization of breeding was designed to support exhibitions, a situation that created an inbuilt and ongoing tension within the chicken-breeding world as it evolved in both Britain and North America – a tension that led to a question which had nothing to do with actual breeding methods: namely, did breeding for fancy mean the same thing as breeding for utility? The use of the show ring as a means of improving other livestock for utility purposes was widespread in both Britain and North America, and similar debates emerged in relation to horse, cattle, and even dog shows.[15] But the beauty/use dichotomy became most blatant within the poultry-breeding/show system.

Concern with beauty only had come by the 1870s to so dominate important early British breeders that many no longer gave lip service to agricultural usefulness.[16] Farm experts in Britain blamed the rising obsession with beauty for an ongoing ruination of older, good producing lines.[17] They condemned the exhibition system, arguing that it had been a terrible mistake and should be abandoned.[18] Beauty breeding became completely detached from utility breeding, and by the 1890s British utility breeders had established their own organization to promote better egg production. They also imported American breeds to work with. Somewhat ironically, British breeders favoured using the working breeds that had been created in North America under an exhibition system that had originated in Britain. Only one significant utility variety was created in Britain, the Cornish (bred in Cornwall from English Game and Indian Game), but even that breed would undergo its most important development in the United States. When British breeders focused on productivity, however, their stock was valuable to breeders in North America. Utility-bred lines thus were crossing the Atlantic both ways by the twentieth century.[19]

The beauty/use dichotomy arose within the North American chicken world as early as breeding for exhibition took hold. An overemphasis on beauty quite quickly led to the Hen Fever, which arose in the eastern seaboard states between 1849 and 1855. Birds sold for fantastic prices, as much as $700 a pair.[20] This demonstration of excessive devotion to beauty and to sport horrified many agriculturalists, who argued that exhibition poultry were supposed to serve the commercial industry by generating useful birds. 'What a price to pay for a hen and rooster!' exclaimed the *Farmer's Advocate* in Canada some years after the events.[21] Despite their disgust for the Hen Fever, Canadian farm experts continued to think that poultry exhibitions, if beauty concerns stayed at a controlled level, could bring about large-scale improvement in the birds. These people thought the situation had corrected itself by the late 1860s in the United States: utility was now served as much as beauty

5 The Cornish, an upgraded Old Fighting Game. The Cornish played a major role in the rise of the broiler industry. Note the extent of meat in the legs compared to the Plymouth Rock. (*American Poultry Journal*, December 1913, 1521)

by the show system. Canadians should pay attention to what was happening in the United States, many believed. In 1868 the *Canada Farmer* explained developing American poultry trends to Canadian readers:

> Poultry exhibitions, it appears, are coming into fashion on this continent. Our neighbors in New York have recently held one in connection with the newly formed Poultry Association, and in other places the example is being followed. We hail the sign with much satisfaction, for we hold that these societies and exhibitions are of no inconsiderable value. No other proof of their utility need be given than the fact of the great improvement in market as well as fancy poultry in England since 1848 and although amusement and fancy may be in the first instance the great incentives, the end attained is general usefulness, and in many cases profit. Markets are better supplied and more birds are kept in farms and elsewhere, establishing an abundance of wholesome and cheap food, to say nothing of the supply of feathers.

Canadians should follow the American example and have more shows for chickens. Even though these would be expensive, the journal stated, they needed to be well organized under the ruling of the British system.[22] A widespread show system in Ontario would trigger considerable growth of the poultry industry, the *Farmer's Advocate* explained. The increased demand for eggs and poultry meat meant that it was time to develop better poultry, and shows promoted this trend, the journal argued.[23] More poultry shows would probably help in the spread of improved stock, agreed the *Canadian Poultry Chronicle*.[24] Buy poultry from fancy breeders, the *Chronicle* advised. It doesn't cost more to keep good poultry over bad.[25] Exhibition breeders in Ontario believed much the same thing. When breeders of show poultry got together in the late 1860s, they were more likely to discuss utility issues than fancy concerns. They clearly saw their work as promoting better agriculture, not simply supporting a sport hobby. The president of the Ontario Poultry Association assessed various breeds for capacity in egg laying, feed conversion, disposition, and general hardiness.[26]

American breeders also believed that the problem of overemphasis on beauty corrected itself after the effects of the Hen Fever in breeding/showing circles had died down. An important American breeder, U.R. Fishel of Indiana, explained the situation by describing how the White Plymouth Rock fit into the show and farm worlds. 'Not only has [the White Rock] been bred for fancy points, but the utility part of the breed has been retained and carefully looked after until today the breed ranks first as egg producers, while as a market fowl there is no breed to compare with them,' he stated. Fishel had bred fancy poultry for many years and had dealt with most known varieties before he started breeding White Rocks. He emphasized his success in producing top-quality White Rocks with both fancy and utility features. 'I remember when the U.R. Fishel White Plymouth Rocks were first introduced,' he reminisced just after 1900; 'people realized that I was striving hard to give them a fancy fowl that was also a business fowl. My combining both beauty and utility in my strain of White Plymouth Rocks has convinced the poultrymen the world over that it pays to breed a variety of fowls that can fill any place where a chicken is wanted.' He added that he had exported stock around the world.[27] As late as the 1920s, Fishel repeated that his concern was for beauty and utility, and added that he had succeeded in making his stock both more beautiful and more useful.[28] American and Canadian breeders, like Fishel, focused on sales in three markets – birds for exhibition, birds not good enough for exhibition to be sold to farmers for egg-production purposes, and birds culled for meat.[29] Ads for breeding stock, however, emphasized that birds had been bred to standards relating to both beauty and utility.[30]

By the late 1880s, in spite of on-going claims of breeders like Fishel that beauty and use were combined in exhibition breeding, the idea that shows successfully encouraged the rise of better farm chickens no longer attracted the same acceptance by the general public that it had by the late 1860s. People increasingly questioned views held by breeders like Fishel, and wondered as well about the effects of the show system on agricultural interests. The Mark Lane *Express* in the United States spoke out strongly about the failure of the breeding/show system to promote useful farm birds in 1888. 'The fancier who minces the matter, preferring to allow the world to continue to believe that exhibitions instruct and improve the people in a particular direction is insincere. In answer to the question, What has the poultry fancy done for profitable poultry? we must answer, clearly enough, nothing.'[31] The *Canadian Poultry Review* agreed with the position taken by the *Express*. Poultry exhibitions were designed to help farmers, not fanciers, an article pointed out in 1892, but since farmers rarely went to shows, they learned nothing about the work of professional breeders. But the writer was not so sure that the exhibition system should be blamed for this state of affairs, evident as it might be, adding that 'too much praise cannot be given to those who come under the name of 'the Fancy Breeder' for the interest

6 An ad for the breeder U.R. Fishel, who promoted his White Plymouth Rocks on the basis of both utility and beauty. The compact body also displays good symmetry and balance. (*American Poultry Journal*, March 1907, 267)

and energy they have taken in improving and bringing to the exhibition standard the various breeds of fowl.'[32]

The apparent focus on beauty made the funding of poultry shows by government increasingly contentious in both Canada and the United States. 'Feathers count and *feathers only*,' the *Farmer's Advocate* fumed in 1900, proclaiming they were the chief object of shows. Look at such breeds as the 'football-haired Polish,' the journal pointed out. Fanciers produced breeds that were 'a curse to any farmer or practical poultryman,' concluded the *Advocate*. Nonsense, responded a fancy breeder, who elaborated by stating that the exhibition standards provided points for utility factors in American-bred classes, and furthermore, of the one hundred points only thirty were devoted to feathering characteristics. He added that one should remember that it was the fancier, not the farmer, who created all the utility breeds in use.[33]

John Dryden, Ontario minister of agriculture, explained to breeders that the support of shows was meant to encourage a better understanding among farmers of what constituted superior poultry. Addressing the Poultry Association of Ontario in 1894, he told members that the government did not fund shows in order simply to provide breeders with prizes. The government hoped that farmers would learn from shows where to go for superior breeding, and the breeders should be alert as to what farmers wanted in poultry. Farmers did not see the colour of a feather as evidence of superior breeding. Utility was what they wanted. Dryden also believed that fanciers did little to help farmers understand how to breed or care for poultry in a more satisfactory way.[34] The thorny problem of state involvement with prize money for exhibition fowl that clearly did not supply farms with useful birds bothered Americans too.[35] Government funding of breeder activity which worked off a show system makes it clear, however, that the state looked to the breeders as experts and therefore sought their guidance on how best to support and encourage superior agricultural breeding practices. Even when breeders emphasized sport breeding, governments were reluctant to abandon the idea that the breeders were the experts in the field.

Breeding Strategies of Craft Breeders Interested in Beauty

Beauty alone could drive breeding methodology, but not without controversy. The North American use of a breeding system for exhibition birds, developed by a man named Belton about 1870 in Britain, indi-

7 A modern Bearded Polish, a bird that can only be described as exotic, and indeed 'football'-headed, with feathers. (Photo by Pete Paterson)

cates that trend.[36] In order to achieve the characteristics Belton desired in both male and female stock, he bred one line for producing males and one for females. He did not try to get the correct males and females from one mating. Two lines were developed: one by mating females producing especially good pullets (young hens) to males producing particularly good pullets, and the other by mating females producing good cockerels (young males) to males producing good cockerels. The final correct males and females that resulted from this breeding program, then, were useless as breeding pairs for show stock, thereby creating a sort of biological lock and bringing new meaning to the idea of mating the 'best to the best.' The breeding lines produced the 'best,' but members of one sex in each line were not the 'best' in themselves. This double mating system was adopted by many North American breeders and provoked a lot of controversy in the North American Barred Plymouth Rock world late in the nineteenth century and into the twentieth.

The trouble arose because male and female Barred Rocks were supposed to be identical in colour and shape of barring (or stripes) across the bird's body and wings. Females had a tendency to be darker than the males, due to the sex inheritance mechanism of colour and the barring trait. The only way to achieve identical and proper colour in both male and female was to breed within a double mating system. For the production of correctly coloured pullets, breeders were advised to select females of the perfect colour, and breed males to them that were lighter in colour and therefore not of desirable colour. The male must be lighter than his mate and his barring as perfect as possible. The lighter the breeding male, the lighter would be his pullet offspring. When high-class females had been obtained year after year from such matings, the cockerels that resulted in such breedings, while not perfect from a colour point of view, would be invaluable as pullet producers. To breed correctly coloured final males, breeders should select large females in shades considerably darker than was considered correct. Put the dark females to perfectly coloured males. The resulting females would not be of good colour, but they could be useful for the production of perfectly coloured males.[37] The emphasis on deep barring marks – in fact, through the feathers and down to the skin –increased the need for double mating. Colour, of course, was just one consideration. Size, shape, and other qualities were selected for as well, making such a breeding program demanding.[38]

The practice of keeping two lines to achieve correct colour annoyed many Barred Plymouth Rock fanciers, who argued that one line should be used for correct males and correct females. These people considered double mating to be theoretically undesirable, because male and female Barred Rocks were designed by nature to be a different colour.[39] Some believed that, over time and with careful selection, males and females would be brought closer together in barring characteristics and in underlying colour within one line (but this had not been accomplished by the 1920s, and by then most had come to accept the fact that barring was sex-linked).[40] Many thought that the Barred Rock naturally threw a variation in colour due to the hereditary background of the breed, and furthermore that an overemphasis on colour had led to a sacrifice of good overall shape.[41]

The breeding of Barred Rocks by single or double mating brought out other contentious issues. Single mating was simpler (and, it might be noted, good White Rocks could be bred successfully by the single mating system),[42] thereby attracting more breeders. The method also made it easier for a buyer to perpetuate the breeder's work without needing replacements for the next generation. A number of breeders thought poultry breeding should be governed by such factors. Because the Barred Rock was popular with farmers, the general public should also be able to regenerate Barred Rocks from stock they owned and not be forced back to the breeder at every generation, it was argued. Furthermore, breeding Barred Rocks should not be an elitist occupation. Others saw the fact that double mating was difficult as an advantage: it tested the mettle of a breeder's ability. The method drew the best breeders (namely, men who did not breed primarily for money, but did so instead to learn more about the skill of breeding) to the Barred Rock, advocates of double mating argued. Because the creation of good Barred Rocks via double mating required considerable breeder ability, supporters claimed that the activity taught people how to breed in a superior manner.[43] It was lessons in breeding, not so much the results of breeding, that made the double mating system important for defenders of the method. Attempts to put an end to the conflict between the two factions revolved around the idea of having two standards, one for light and one for dark birds.[44] More people would breed Barred Rocks under these conditions, many believed. When the Standard of Perfection came up for revision yet again in the mid-1920s, the battle lines were in place, as they had been since at least 1897. No solution to the problem was in sight.[45]

The Barred Rock breeding history elucidates features of the early poultry-breeding world. First, it is clear that beauty, completely separate from utility factors, could direct how breeding proceeded. Feather colouring did not relate to either egg or meat productivity. More noteworthy than this beauty emphasis in breeding, however, is the fact that the use of double mating implies that chicken breeders, as early as the nineteenth century, understood breeding schemes that could bring about biological locks. Equally noteworthy is evidence that the breeders' real problem with the double-mating method revolved, not around its application for beauty breeding, but rather around the idea of biological locks. Following a method that dictated locks of this nature did not make sense to them. Breeders who opposed double mating questioned the desirability of forcing farmers back to breeders for replacements, which the biological lock imposed by the double mating system brought about.

Problems in Establishing Breed Standards

Committees of the American Poultry Association decided on how to revise standards for each variety if that need arose. New breeds could also be admitted if the American Poultry Association was convinced that type bred reasonably truly, with respect to body shape, wing and leg structure, overall balance, and feathering. Many important aspects of the standard were not intended, however, to relate to beauty. Body shape, believed to indicate both egg-laying ability and the general vigour of the individual, played a large role in the standard's rulings. Heated discussions over shape often took place, and disagreements arose over the truthfulness of illustrations designed to aid one's understanding of what the standard's wording meant per breed. The first edition of the Standard of Perfection to carry illustrations came out in 1883, and it was so disliked that it was declared 'obsolete.' Preparation for new revisions in 1910 and 1911 broke down over conflict around suggested illustrations, forcing the American Poultry Association to revert back to the 1905 version.[46] A new Standard of Perfection was not released until 1915.

Imagery, in spite of problems around it, played an important role in poultry breeding. It was believed that it was possible to create pictures that illustrated and therefore explained correctness of body shape, vital to both productivity and beauty. The problem of making an image match what all poultrymen believed a variety should look like, particularly as to body shape, proved over the years to be even harder than to find words to express perfection, as the endless changes to the Barred

STANDARD PLYMOUTH ROCK SHAPE—MALE.

"A Composite Ideal From Live Models"—As First Submitted by the Reliable Poultry Journal for Criticisms of Judges and Breeders.

8(a) Illustration first suggested by the *Reliable Poultry Journal* to depict the ideal shape of the Male Barred Plymouth Rock, a large, thick bird that also could compete with Leghorns in the ability to lay eggs at this time. (Reliable Poultry Journal, *The Plymouth Rocks, Barred, White and Buff* [Quincy, IL: Reliable Poultry Journal Publishing Co., 1899], 13)

STANDARD PLYMOUTH ROCK SHAPE—MALE.

As Corrected to Meet the Criticisms of Judges and Breeders.

8(b) Illustration proposed after revisions suggested by breeders and judges for the ideal shape of the Male Barred Plymouth Rock. Note how minor the changes are, and how much they relate to perceived balance. Both models show a strong meat tendency in the compact and large body. (*The Plymouth Rocks*, 17)

STANDARD PLYMOUTH ROCK SHAPE—FEMALE.

"A Composite Ideal From Live Models"—As First Submitted by the Reliable Poultry Journal for Criticisms of Judges and Breeders.

9(a): Illustration first suggested by the *Reliable Poultry Journal* to depict the ideal shape of the Female Barred Plymouth Rock. There is not much deviation in shape between the sexes in the Rocks. The female is slightly smaller, but she does demonstrate the same thick, compact form. (*The Plymouth Rocks*, 19)

STANDARD PLYMOUTH ROCK SHAPE—FEMALE.
As Corrected to Meet the Criticisms of Judges and Breeders.

9(b): Illustration proposed after revisions suggested by breeders and judges for the ideal shape of the Female Barred Plymouth Rock. As with the male, alterations are so minor that they reflect only an individual judgment of relative balance. (*The Plymouth Rocks*, 21)

Plymouth Rock images clearly showed in the 1897–9 period. Late in 1896 the *Reliable Poultry Journal* arranged for F.L. Sewell, a noted poultry artist, to execute a drawing of the perfect male and female Barred Plymouth Rock as set out by the Standard of Perfection. The journal asked for comments from judges and breeders as to the correctness of the images.[47] A flood of responses came in, many criticizing the images on a variety of points relating to such issues as length of legs and neck, barring, and overall shape. Attitudes to the all-important feature of general body shape (in both male and female), however, seemed to be the most contentious. The male brought out particularly conflicting opinions. Of note are the views of two distinguished judges and breeders. The great general breeder and judge I.K. Felch believed the representation to be inaccurate, and he trimmed the image with his scissors to make it clear what he thought should be altered to represent correct body shape.[48] A.C. Hawkins, a judge and breeder specifically of Barred and White Plymouth Rocks, did not share Felch's opinion in this matter of male body shape. He wrote: 'Concerning the cut of the male bird, I think it the best in form I ever saw, and I would not change it in any way. Mr. Sewell has done himself much credit, and the American Barred Plymouth Rock Club can do no better than take it for a standard for form.'[49] When the numerous suggestions for change, or lack of change in all characters – body shape, wing size, feather markings, and so on – had been collected, Sewell re-submitted new images for both male and female which incorporated the various revisions as best he could.

In 1901 the *Reliable Poultry Journal* undertook another such study, this time in relation to the perfect Leghorn, and hired Sewell to do the same job of drawing and redrawing.[50] While there seemed to be less contention over correctness than had been the case with the Rocks, that did not mean everyone agreed on shape and feathering as another set of drawings of Leghorns revealed. In 1914 L.A. Stahmer undertook to illustrate the perfect Leghorn shape and asked for the opinions of breeders. Many replies came in modifying the shape. Stahmer redid the image and the *American Poultry Journal* printed it along with the artist's commentary on the procedure.[51] It is interesting to compare the Leghorn perfection images of 1901 and 1914. Evidently, beauty styles could change quite rapidly, as the differing tail carriage makes clear.

Because bird shape was important for both productivity and beauty, it played a role in sales. As a result, it became common to use untruthful illustrations of birds in breeder advertisements. The problem of fal-

S. C. WHITE LEGHORNS AS BRED BY D. W. YOUNG, RIDGEWOOD, N. J.

10 Illustration suggesting Perfect White Leghorns in 1901. These birds are smaller than the Rocks and carry ornate feathering in contrast to the heavier, meatier birds. There is also greater sex deviation between the two. The Leghorn hen's comb flops over her head, a characteristic that resulted in the large floppy heads of women being called 'Leghorns.' (Reliable Poultry Journal, *The Leghorns: Brown, White, Black, Buff and Duckwing* [Quincy: IL: Reliable Poultry Journal Publishing Co., 1901], 58)

11 Illustration suggesting Perfect White Leghorns in 1914. Note the rather dramatic change in the tail carriage, particularly in the male. This characteristic surely had little to do with productivity. But clearly, perceptions of beauty, or desired style, could change over a relatively short period. (*American Poultry Journal*, February 1914, inset)

sified or retouched advertising images worried people. In 1915 Felch addressed this problem, saying: 'The sooner purchasers are made satisfied with good, true, unretouched photographs of living specimens ... there will be much better understanding between breeders and their customers. Illustrations should show fowls as they appear in the breeders' yards.' The editor of the *American Poultry Journal* could not have agreed more, and stated: 'We have been subjected to a long, long siege of idealized pictures, retouched photos, or faked photographs (as you please) of Standard bred fowls. Almost everyone is doing it. The pictures showed us wonderful visions of feathered perfection, the like of which never existed except in the imagination of the breeder and the artist he employed.'[52]

The Rise of a Commercial Egg Industry

The establishment of breeds and breeding organizations took place within a rising commercial market for poultry products, eggs in particular, in North America. The new emphasis on eggs, not poultry meat, presented a significant change, historically speaking. Utility chicken had been used more for meat than egg purposes (although eggs too were consumed) in both the Old and New World before the advent of the nineteenth century. In the United States, cheap grain in the Ohio valley and better transportation methods encouraged the raising of poultry for commercial egg production in the Midwest between 1825 and 1860. By 1863 egg shipments reached New York City from Ohio, Indiana, Illinois, and Minnesota. By 1874 eggs came to the city from as far away as Mexico and Canada.[53] The commercial egg industry in Canada took shape somewhat later. Chickens had been brought to Canada as early as European settlement took place and stock subsequently moved freely between that country and the United States. Even so, contemporary poultry experts did not think a viable commercial industry for poultry products existed in Canada (in spite of evidence of some exportation to the United States) until the early 1880s, when a considerable demand for fresh eggs developed in urban centres such as Toronto and Montreal.[54]

The first US census of poultry in 1880 showed that there were 105 million chickens in the United States. Growth over the next ten years showed an increase of 153 per cent. By 1910 the numbers had steadied.[55] Still, 88 per cent of American farms raised chickens.[56] In contrast, the first major Canadian growth spurt in chicken numbers took place

12 An example of artistic licence used in poultry imagery for advertising reasons. A 'Fishel' White Plymouth Rock Cockerel. The bird strikes a rather unnatural pose. Feathers are so smooth they are invisible. (*American Poultry Journal*, December 1907, n.p.)

twenty years later, reflecting Canada's delayed development when compared to that of the United States. The first Canadian census for poultry (which included turkeys and ducks along with chickens) in 1891 revealed that the nation had 14 million birds). While growth was steady after that time, the first major spurt was between 1901 and 1911, with an increase of 77 per cent over that ten-year period.[57] It might be noted that the category 'chickens' dominated poultry generally. In 1921, for example, chickens represented over 95 per cent of all Canadian poultry. It was estimated in the United States in 1914 that some 94 per cent of poultry were chickens. Chicken raising was common on city and town lots in the United States and Canada, but the vast majority of birds resided on farms.[58] This rapid growth in the poultry population reflected the fact that an ever-expanding market existed for poultry products over the late nineteenth and early twentieth centuries.

Poultry keepers wanted better egg yields generally, but they focused on the laying of winter eggs (hens tended to lay in the spring and early summer only).[59] Experiments done at the Ontario Agricultural College on egg laying revealed clearly that volatility of egg production by hens over the year had a devastating effect on profitability because of the consistency of feed costs. The egg-laying specialist Leghorn might produce as many as sixteen eggs in March, seventeen in April, and thirteen eggs in May, but virtually nothing from September to November.[60] Profits fluctuated violently, as a result, in relation to the number of eggs the hen laid. She had to be fed if she laid or not. Clearly, if a farmer wanted to earn a profit from eggs, winter laying had to increase. Egg colour, however, was important too and the end market influenced what breed a producer would use: white was favoured near New York (from Leghorns), for example, and brown near Boston (from Barred Rocks).[61]

Breeding Strategies of Craft Breeders Interested in the Egg Industry

This emphasis on egg laying resulted in the rise over the nineteenth century of other complicated breeding strategies aimed primarily at egg production. All were invented by professional fancy breeders who believed that utility should not be sacrificed in the pursuit of beauty. One important breeding scheme, set out by the American master breeder I.K. Felch in the 1870s, spanned generations and balanced the benefits of sanguineous breeding against the dangers such practices could incur. Inbreeding preserved and perpetuated desired characteristics, while the expression of undesirable features, which might emerge with excessive inbreeding, was guarded against by following Sebright's percentage system of inbreeding. By recombining the blood of an original breeding pair from different mating combinations over generations of their descendents, one could inbreed (or line breed, according to Sebright's definition of the term) forever without experiencing seriously reduced vigour.[62] The heredity of the original pair could be preserved in various blends. And if the original parents were not closely inbred, the dangers of related matings could be checked even more, by reverting back to the equal input of blood from both of the original pair.[63] This situation did not, of course, mean that all stock carrying the same percentages would be identical. Full brothers and sisters are not all alike, a fact that any sensible breeder learned through experience. But Felch could avoid a lot of the dangers of inbreeding in this fashion, and with careful selection he could shift the hereditary make-up of the flock towards uniformity.

Felch noted that as early as the 1860s good poultry breeders knew that cross breeding improved productivity. 'It makes no difference whether two breeds are crossed or two specimens of a breed of widely different parentage, the progeny will be more prolific than the breeds or families crossed,' he stated. The problem was that the progeny would not breed on truly, Felch warned. Inbreeding created strains that bred truly, he added, but the method introduced the tendency to reduce vigour. By using his breeding chart, Felch maintained, the original vigour of the first crossed generation could be preserved and, if need be, returned to the breeding line. He concluded as follows: 'Your head must be level to mate not but specimens that are strong, healthy, fine in type ... As long as you can create groups representing half of the blood of each of the original Adam and Eve of your flock as reservoirs from which you can draw new blood for your mating in such a way that each group of chicks will show a change in their blood from that of their sires and dams. That is the secret of inbreeding. It may be revolting to apply this rule to the human family; nevertheless by it could a single man and woman populate an isolated island or region with healthy human beings.'[64]

Felch's ideas and charts would be quoted and re-quoted (and sometimes presented with a few modifications that really were 'little more than a steal')[65] throughout his life and after his death in 1918 at the age of 84.[66] More than ten years after his death, Felch's system was still presented as a safe way to practise inbreeding, but by this time in conjunction with even more complicated breeding plans. An article pub-

From Barnyard Scavenger to Bird of Beauty and Use   61

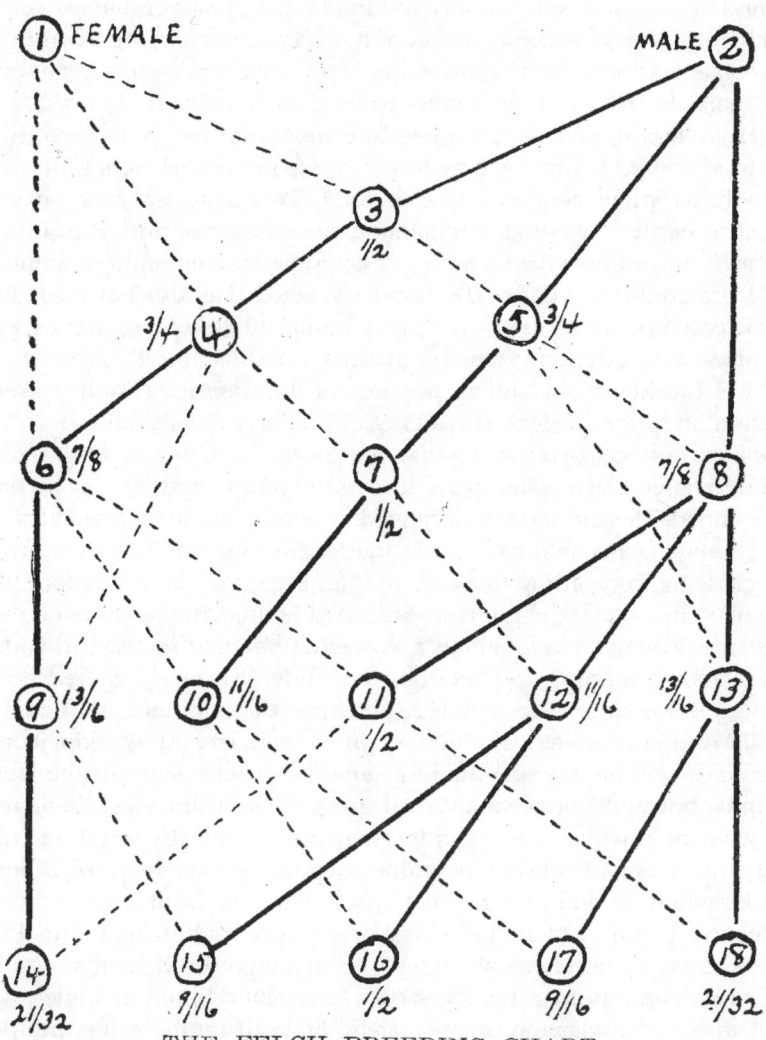

THE FELCH BREEDING CHART.
The plan of line breeding advocated by I. K. Felch, Natick, Mass., the old veteran judge and breeder who for 40 years has been a deep student of the problems of mating.

13 A diagram explaining Felch's famous breeding chart. (*American Poultry Journal*, August 1910, 974)

lished in 1928, for example, advised the keeping of several lines within each strain, and the preservation of those lines as separate for at least three generations. Each separate line within the strain should be bred for males to cross and for females to cross each on the other line.[67] The Felch system was, in fact, the best and most fully explained breeding method offered to poultry breeders from the time of Sebright until well into the twentieth century. As Felch put it, 'We can mix the blood of our birds as easily as we mix paints that give us different tints of colour.'[68] Clearly, he subscribed to a newly developing view in biology, namely, that life could be manipulated and managed. He also believed that birds could be made to breed truly in sufficient volume to make them as readily available as any other 'manufactured' item.

H.H. Stoddard, a founding member of the American Poultry Association and an experienced American breeder, as well as a renowned publisher of poultry journals that emphasized utility breeding, wrote numerous articles explaining a different breeding system that he had developed. The method was designed to create producing inbred flocks of egg-laying hens, but maintain the healthy vigour that arose from outcrossing in breeding flocks. Careful inbreeding could increase, not decrease, the level of egg laying, Stoddard argued, but he believed that the impairment of vigour that followed inbreeding in the end counterbalanced and cancelled better productivity.[69] Therefore, inbred stock should never be used to create breeding lines.[70] Such stock should only be used to produce eggs for market. In order to create a working commercial flock, he advised the breeder to start with sixteen unrelated strains, but within the same breed, from the best lines possible, and slowly combine the genetics of these over five years, by selecting only males from certain lines for breeding and only females from other lines for breeding. Combining the lines would be done through this type of selection. Culling should be done at every level of breeding. The final cross of purely unrelated stock resulted in a vigorous flock that would be inbred, brother to sister, for at least four years. The inbred lines, not the unrelated crossing of lines, were to be used for table-egg production. The stock would be weakened by the inbreeding after four years, but that passage of time would allow the breeder to work up his original sixteen strains to a new final cross, which would bring back vigour for the next group of inbred table-egg layers.[71] Stoddard advised a well-planned balance between inbreeding and outcrossing, just as Felch did, in order to harvest the advantages that inbreeding could bring without incurring its dangers. But Stoddard advised stronger and more intense

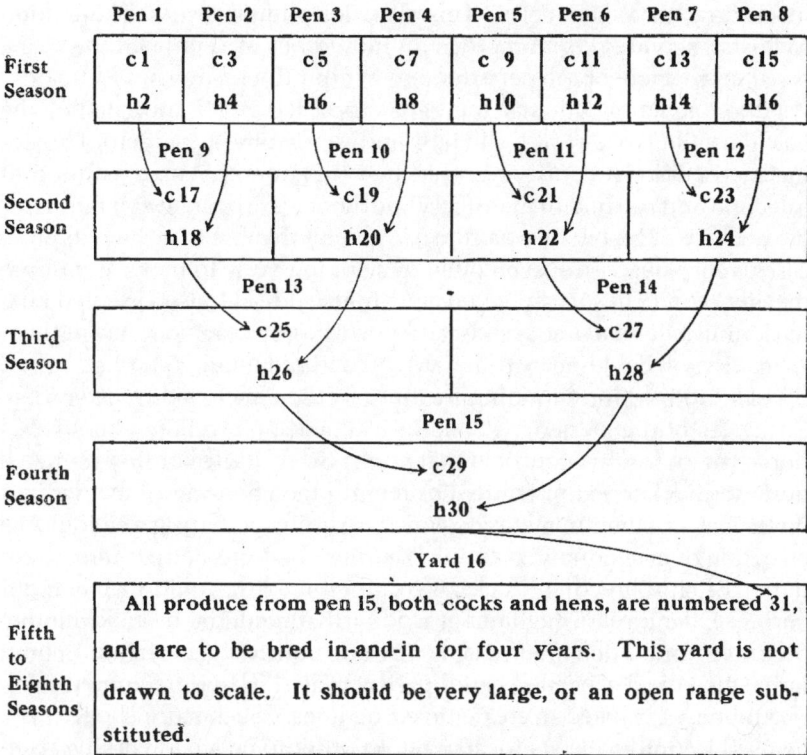

14 A diagram explaining H.H. Stoddard's breeding strategy for utility poultry. (*American Poultry Journal*, December 1913, 1542)

inbreeding than had Felch. This situation required greater attention to the preservation of numerous foundation lines, and from these, the constant renewal of the perpetuating egg-producing lines.

Fundamentally, whether a breeder used the Felch approach or the Stoddard plan (or a strategy that utilized some aspects of each), a breeding methodology always involved two separate processes: method of selection and method of mating. While these were not one and the same, the activity of breeding was complicated by the fact that the two processes interplayed with each other in such a way as to make an infinite myriad of overall strategies possible. Method of selection involved the reasoning behind a breeder's decision to collect certain individuals over others in any breeding population. Method of mating described how a breeder utilized individuals, already selected, in a breeding operation. Examples of methods of selection that could be used are: mass selection – the choosing of individuals on the basis of their ability to match desired characteristics; family selection – the choosing of individuals on the basis of their relatedness within a family; pedigree selection – the choosing of individuals on the basis of their pedigree; or progeny selection – the choosing of individuals on the basis of the quality of their offspring. Examples of methods of mating that could be used within the breeding population are: crossing an inbred individual on a non-inbred individual within a breed; crossing two related individuals; or crossing one breed with another. Examples of how a mating method might follow a selection method are: after a group of animals had been selected for breeding on the basis of their family relatedness, they can either be mated in such a way as to increase relatedness – that is, breed brother to sister – or to downplay as much as possible that relatedness – that is, look for cousin connections. If a group of animals had been selected on the basis of their ability to produce offspring, individuals can be mated to promote relatedness or maintain unrelatedness. There is no clear way to describe how breeding actually proceeded under these conditions, because the many ways that selection method could be combined with breeding method made an infinite variety of approaches possible at any one time. Probably breeders shifted their approach to their breeding operations with changing circumstances.

Breeders and Producers:
Conflicting Theoretical Approaches to Breeding

With the rise of a commercial agricultural industry based on birds, the farm poultry world began to show clear signs of internal divisions. Those interested in producing fowl tended to fall into two separate camps: namely, one of true breeders who created distinct lines of stock and one of producers/growers who simply multiplied or used the birds. The two groups held differing views on what was desired in the birds. The breeders wanted fowl which could perpetuate themselves within different pure breeds as defined by the Standard of Perfection. The producer/growers were less concerned with accurate perpetuation of type than with productivity of stock on the farm, and while they might seek to regenerate stock from what they owned for a short time, they were always prepared to bring in fresh blood. This deviation meant the two groups did not necessarily share approaches to breeding methodology, as to either the process of selection or that of mating. Breeders relied on some form of inbreeding which they defined as line breeding. Producer/growers strongly resisted inbreeding and were inclined to favour cross-breeding for better vigour. Over the 1870s the North American farm and poultry press explained how the breeders used what appeared to be controlled inbreeding in considerable detail for producer/growers.[72] References to the inbreeding practices of the breeders made many producer/growers simply avoid the pure breeds. Before 1900 the producer/growers often preferred to work with birds that resulted from various crosses of breeds.

After 1900 producer/growers were more likely to favour using the single breeds, but they still resisted inbreeding. The move away from cross-breeding by producer/growers of table eggs made the two groups clash even more directly over breeder inbreeding techniques. Since there was no even dividing line between the two poultry occupations (breeders were often producers too, but most producers were not breeders), what chicken breeders and producer/growers did (and should do), however, could become confused in the minds of contemporary agricultural experts, a fact which helped both to mask the beauty/utility problem in breeding and make the very meaning of the word 'breeding' in poultry affairs confusing.

Cross-breeding could be done to increase better egg or meat yields, but it was most commonly associated throughout the nineteenth century with breeding for meat. Even though the egg industry received the most attention from producer/growers, considerable interest in meat breeding still existed. Double cross mating systems based on three breeds might be followed to produce a good table fowl. Stabilizing a line by this method, however, was clearly more difficult.[73] Crossing and re-crossing stock might shift features or characters, but the method did not fix that changed type. Many producer/growers, however, were not concerned with fixed type. The agricultural press indicated that North American poultry keepers, even if they did not undertake complicated cross-breeding programs, commonly cross-bred birds for meat purposes until the end of the nineteenth century by introducing fresh blood of another breed, in the belief that greater hardiness would result, as well as faster growth and better fecundity.[74] *Farming*, a Canadian journal, noted in 1898 the propensity to seek this type of increased vigour. The crossing of two breeds of fowl worked well for the production of superior birds, the journal had to admit. 'By crossing two ... breeds that are very dissimilar, we secure an increase in hardiness in the first cross, as well as the special qualities in each breed to a high degree.'[75]

The journals condemned cross-breeding, even if it led to increased vigour and better productivity. 'There is, it must be confessed, one great disadvantage attending to the rearing of cross-bred fowls – they are ... unsaleable as stock birds,' the *Canadian Poultry Chronicle* warned as early as 1871.[76] Crossing of breeds also seemed to be out of line with good farming practices for the agricultural experts. Breeds that carried good quality and also bred truly for such quality had been established.[77] 'To attempt to produce a bird of merit by crossing would be simply producing something that can be found among the pure breeds already,' the *Farmer's Advocate* explained in 1895. 'In fact, such work [would] be but a repetition of what has already been done, and it is in one respect a waste of time. Anything that is wanted can be procured from the pure breeds.'[78] One thoughtful person, writing to the *American Poultry Journal* in 1908, agreed. 'Every little while someone tells us that cross bred hens do better than pure bred, so they will take a good flock of pure bred hens and mate them to a male of another breed, and so spoil the whole lot. Spoil them? Well perhaps not entirely ..., [but] nothing is gained by crossing different breeds of poultry, but much is lost by doing so.'[79] As the egg industry gathered increasing momentum early in the twentieth century, any attempt to focus on meat breeding (and concurrently on cross breeding) decreased significantly. When it came to discussing the desirability of cross-breeding, farm journals evidently mixed the separate interests of breeder and producer/grower in a confused way.

Obsession with purity drove the ever-increasing emphasis of agricultural experts on using a single breed, but that obsession had become complicated by the late nineteenth century in North America. Purity often meant two overlapping things to agricultural experts: lack of deterioration or degeneration, and repeatability in the breeding lines. The crossing of animal breeds (often described as hybridizing) provoked an array of conflicting attitudes from agricultural experts at the end of the nineteenth and beginning of the twentieth centuries concerning the problem of degeneration. While opinions held by agricultural experts that might be described as 'animal eugenics' supported the thought that cross-breeding led to race deterioration and the advent of 'mongrels,'[80] it would be too simple to suggest that these people believed the concept of 'purity' was important simply because it counteracted the idea of 'mongrels.' The fact that such 'purity' guaranteed stock so bred would breed truly – and therefore could reproduce in its offspring the good qualities that it possessed – was at the heart of the matter. Purity in breeding meant ensured repeatability; lack of purity meant unknown results.

In spite of these purity problems, support of inbreeding showed signs of weakening within large livestock circles in North America. While still often touted as the way to produce good stock, Bakewellian inbreeding principles were less likely to be followed. Bakewell's system had been designed to create 'breeds.' In the eighteenth century and early nineteenth century, livestock improvement was defined as the fixing of breeds, namely, the extension of good qualities normally found only in a few individuals to larger numbers. Most modern breeds of livestock developed between the mid-nineteenth century and the early twentieth century. Livestock breeders tended to avoid sanguineous matings in any established breed, because type could be maintained without using that strategy.[81] Increasingly, distaste for 'incestuous' breeding became common. The very ethos behind purebred breeding, that is, the desire for or recognition of improvement, shifted. Purebred breeding came to be seen as a method to preserve type, not improve it. It has been argued that the very idea of 'preservation' rather than 'improvement' was an outcome of the North American desire to import stock already developed to type, rather than breed for better lines of horses, cattle, swine, and sheep. If various breeds had been built up in North America, the global allegiance to the purebred system as a preserver of the status quo might not have been so strong.[82] The tendency of North Americans to import livestock breeds from other countries, and then to simply perpetuate their good qualities served to retard experimentation.[83] Under these conditions, any form of inbreeding increasingly looked less appealing.

Fear of reduced vigour from inbreeding increasingly concerned poultrymen by the late nineteenth century, at a time when most of the important chicken breeds and varieties had become established. A move, similar to that in large livestock circles, away from sanguineous matings appeared among chicken breeders.[84] Even so, not all breeders rejected blood-related breeding systems (in fact, the most prominent ones like Felch did not). But the fact that some resisted inbreeding made the issue contentious and also divisive within breeding circles. Breeder concerns over inbreeding reverberated in the world of producer/growers, strengthening opposition in those quarters to any form of inbreeding in stock.

There were other issues which made producer/growers leery of the fancy breeders. The press worried, for example, about the dishonest dealings of some breeders. Fancy-priced eggs often did not hatch or else brought forth what were clearly cross-bred chicks. Prices and prize winning were no guarantee of fair dealing, the *Farmer's Advocate* told its readers in 1870.[85] Sharp dealings seemed to encourage farmers to turn away from fancy breeding.[86] An article in 1883 admitted fraudulence, but reminded its readers that not all breeders were shady in their dealings.[87]

The Predominance of Men in the Breeding Sector and Women in the Producing Sector: A Gender Cleavage

The breeder and producer/grower divide was further complicated by the gender of the people involved. Most breeders were men who controlled the breeding of birds, while women raised the breeder's stock in order to produce marketable eggs and meat. There does not seem to be any disagreement in contemporary literature on the nature of women's role in the commercial industry: they made up the vast majority of producer/growers.[88] Ads for poultry feed and equipment also seemed to be aimed at women.[89] Articles referring to actual breeding strategies for farm poultry seemed almost entirely to have been submitted by men (either fancy- or utility-oriented breeders) to the press or to have been written by press editors. Women wrote articles in the farm and poultry press, but rarely on the subject of breeding, although preference for breed was frequently stated.[90]

The female domination of the production industry was perceived as making it backward, or underdeveloped in the eyes of many, and thus it was difficult for the birds to shed their reputation as lowly stock.

Women could be decidedly sentimental about their birds. Female-written stories usually carried a note of interest, compassion, and even love for the birds. Chicks were referred to as 'little fellows,' 'little biddies,' and 'downy little chickens that [were] so cute.'[91] One farm woman wrote that she liked to let a mother hen enjoy her baby chicks, and did not send the little ones to a surrogate mother hen in order to encourage the real mother to lay again and faster.[92] Another stated that she named all her hens, and added, 'The delight and interest I take in all these mothers and their fascinating babies is worth many dollars to me.'[93]

Sentimentality did nothing to make men in general take poultry seriously, but chickens and women were perceived by the masculine world to have a natural affinity to each other. In 1882 the *Farmer's Advocate* carried an article entitled 'Poultry Keeping for Women and Children.' 'Properly managed the keeping of poultry can be made profitable, and it should have a place among the industries of the farm,' the journal stated, and elaborated as follows:

> From fifty to a hundred hens might be kept on many farms, at small expense, and a considerable income be derived from them. The care of them should be entrusted to children and women. All the work required is of a light and pleasant kind, and a child of ten or twelve years could do most that would be required ... The care of poultry is also suitable for women. Many a frail woman who devotes all her energies to household work and cares, would find that something to call her out doors several times a day and divert her attention, would soon effect a remarkable change in her feelings and appearance ... Some women who find it necessary to increase their scanty income by 'taking in work,' might find the keeping of a flock of good hens an easier way of attaining the object desired.[94]

The themes of female poverty and poultry, the healthfulness of poultry work for women because of their frailty, and the innate femininity of the occupation underlay sentiments in farm-press articles with frequency.[95] Yet, at the same time that the press extolled the virtues of poultry keeping for women, advice on how to raise poultry for eggs or meat was normally aimed at the 'farmer,' not the farm wife. Perhaps one of the most surprising aspects of gender and poultry keeping, as exhibited in either the poultry or farm press, was the constant false 'masculinizing' of the producer/grower.[96] Anyone offering advice on how to operate in the commercial industry addressed the 'farmer,' but rarely the true producer/grower, namely, the farm wife.[97]

Open masculine disdain for poultry raising and work because of its feminine basis could also be found in the North American farm press. 'Farmers seem to think poultry are all very well for the women, and keep a few to please them; but they do not regard them as part of their farming operations, and worthy of the same consideration as they give their sheep, cattle and swine,' the *Farmer's Advocate*, for example, noted in 1877.[98] In 1883 the *Advocate* quoted an American article on the subject.

> The extent and importance to which the interest has grown is almost incredible: especially so to those who remember how, in their boyhood days, the chickens were looked upon as a necessary nuisance, to be tolerated because the female portion of the household looked to them for a supply of pin money. In those days – and those to us by-gone days are still to be found exemplified in many sections of the country – the fowls were regarded as a species of freebooters, living by their wits and preying upon the industry of the men-folks.[99]

The poultry industry continued well into the twentieth century to be seen as a secondary livestock industry due to the role of women in it. The reluctance of males to take poultry keeping seriously angered the *Globe* in Canada. '"Women's Work," So-Called, is Source of Farm Profit' ... Work is no longer pin-money,' the newspaper proclaimed in 1915.

> With that airy disdain of so-called trifles so commonly found in the 'male of the species,' the man on the average Ontario farm has little or no interest in the fowl his farm maintains. The flock may be pure-bred or pure mongrel; it may be large or small; it may be a revenue producer or a standing debt – to him it is just 'the hens,' and quite below his notice ... The lord of the manor has no place in his thoughts for the problems of poultrydom. Indeed he will condescendingly inform you that he doesn't know how many fowl are kept, that the women look after them ... Working out the case one step further, we are safe in asserting that the bulk of dressed poultry put before the great consuming market of the United States and Ontario at points in this Province is in no way dependent upon the business carried on by large commercial poultry plants. It results almost entirely from the unspecialized, inexpert, too-often discounted work of the women on the ordinary small farms of the Province. The farm women do the work; the agricultural industry – man-made, of course – gets the credit.[100]

Letters written by women confirm masculine disdain for poultry and a male unwillingness to finance poultry keeping to any degree. At a Women's Institute meeting in 1912, for example, one Canadian farm wife advised those listening as follows: 'One of the wisest things to do when engaged in the poultry business is to get the man of the house interested with you, for there are many things about the poultry-yard in which you will need his helping hand' and once he is interested he won't say hens eat too much.[101] Another woman described her work with pure Barred Rocks, how she kept track of good egg layers, and how this in turn affected her husband. She started with birds 'all of reputed excellent laying strain,' leg-marked the pullets, and observed egg laying. She culled the poor ones, and noted her feed costs and income. 'In conclusion I would like you to know that this year my husband who was so skeptical about poultry profits five years ago, is prepared to build an up-to-date poultry house to accommodate one hundred and fifty hens, and that I have absolutely no complaints forthcoming when I need more hen feed or when I ask him to clear out the hen house; in fact the suggestion is usually his own.'[102] American women also tried to convince their husbands to help by building hen houses, and thereby convinced men that poultry affairs should be taken seriously. But even after women of the South began to emphasize poultry production, and in spite of masculine help if it actually materialized, many observers noted that a male lack of respect for poultry affairs persisted.[103]

The economic value of the poultry industry remains more hidden than that of the male-oriented wheat, corn, beef cattle, or hog industries because of lack of appropriate data. Devaluation of poultry keeping, because of its feminized base, probably partially explains why more detailed information on chicken-farming patterns in the nineteenth and first part of the twentieth century does not exist. So does the fact that much of poultry farming was directed at home consumption, making that aspect of its economic value hard to calculate. The situation also made chicken farming appear to be poorly commercialized.[104] While the importance of home consumption to production is hard to estimate with any accuracy, we can obtain a rough picture from various sources. In 1909, for example, one woman in Ontario reported her records revealed that the family consumed roughly 26 per cent of the eggs she produced.[105] In 1916 a bulletin released by the US Department of Agriculture confirmed the same pattern in the United States: chickens played an important role in home consumption.[106] It has been estimated that in 1929, more than 50 per cent of poultry raised and eggs produced in North Carolina were reserved for home use.[107]

The strength of both home consumption and local bartering of eggs can also be deduced from the egg-laying characteristics of chicken breeds that were predominantly used by women. The popularity of the dual-purpose birds in the late nineteenth and first part of the twentieth century (the Barred Rock in particular) implied considerable production and consumption of brown eggs, because the Rocks lay only brown eggs. But these had a very restricted geographic commercial market area. The general market preference in North America outside New England was for white eggs. Poultry keepers living in areas like New England were clearly encouraged to keep Rocks, but Barred Rocks were exceedingly popular in other regions; Ontario, for example. Although it is true that Ontario eggs could be shipped to New England,[108] this pattern does not seem to explain adequately the high level of brown-egg production. It is not logical to assume that Ontario's Barred Rock eggs all ended up in New England, or that large urban centres like Toronto and Montreal consumed only brown eggs. Brown eggs must have served farm families with food in a substantial way throughout North America, and also been part of a localized but hidden barter trade.

Aside from home consumption or barter trade, eggs contributed significantly to farm income. A study of one Ohio family's revenue in 1888 revealed that poultry supplied at least 10 to 20 per cent of the farm's income.[109] Another American study done between 1912 and 1916 indicated that the average distribution of receipts for poultry products over the five years represented 19 per cent of farm income, and in certain areas as much as 23 per cent. The study also found that the sale of poultry and eggs constituted the second highest source of income.[110] Reviews of farm work in the American South reveal similar patterns early in the twentieth century. Women depended on poultry for income, and for the ability to trade in other commodities. It is evident that income from poultry in this region made a substantial difference to the family's budget.[111] The home-consumption and barter-trade aspects of poultry farming continued for a considerable length of time to make many farmers not take it seriously. Men must cease to think of the poultry industry as one generating pin money, the *Farmer's Advocate* argued as late as 1949. 'The pin money era is gone for poultry producers,' the journal stated.[112]

Regardless of the inbreeding/cross-breeding/purity conflict, the beauty/utility issue, business ethical problems, and the gender divide between breeder and grower/producer, fancy breeders continued to make up the breeding arm of the commercial egg industry. They still supplied the basic seed stock used by farm wives. Increasingly the breeders saw themselves as the leaders of the poultry industry.[113] They basked in their self-importance as the breeding arm in the developing commercial poultry industry, but continued to wonder why poultry keepers did not seem to appreciate their birds more. Producer/growers did not seem interested in helping themselves by looking to advanced forms of livestock breeding. 'Why people will continue to keep a lot of uncouth mongrels possessing neither intrinsic worth nor outward beauty, is one of the conundrums that no breeder of choice fowls can solve,' the *Canadian Poultry Review* said heatedly in 1892. 'Most people who keep such wretched specimens of poultry, will tell you "they don't pay" and this is to be presumed in their reason for not getting something better,' the journal explained.[114] As early as the 1870s it was common for people keeping chickens to find that birds coming from exhibition lines often lacked the hardiness of farmyard birds. Various reasons were put forward to explain this awkward situation. One was that farmers or their wives did not feed or house stock in the same way as did the fanciers.[115] Regardless, fanciers saw themselves as the producers of seed stock, and lack of producer/grower appreciation for that fact continued to annoy them. 'There is no getting around the fact that the fanciers of today, as a class of stock breeders and perfectors, are receiving too little for their efforts,' stated the *American Poultry Journal* in 1908.[116] 'The average farmer or poultry keeper realizes but little of the valuation of a flock of thoroughbred poultry, as he has the conception to believe they are for the purpose of having show and fancy stock,' the journal reiterated in 1910.[117]

By 1900 a commercial poultry industry with an emphasis on egg production had taken hold in North America. In spite of deeply divisive issues between breeders and producer/growers, the exhibition breeding industry held hegemony over how breeding should proceed. It was supported by a sophisticated organization of breeders who had developed complex breeding methods directed at both beauty and productivity, and a complete social structure to protect this breeding work. Something new would shortly enter the picture: namely, genetics.

Chapter Three

The Development of Agricultural Genetics in Relation to North American Chicken Breeding

An all-encompassing concern with improved farming developed over the nineteenth century and continued into the twentieth in North America. Better crop husbandry practices involving rotation systems, land drainage schemes, and efforts to understand the nature of soil chemistry which supported the growth of plants all attracted the attention of scientists and agriculturalists in the first half of the nineteenth century.[1] Governments actively encouraged the pursuit of research in these areas and soon established state organizations to extend that work. By the second half of the nineteenth century both the United States and Canada had a central governmental body dedicated to various interests of agriculture, and by late in the century government involvement in farming had expanded enormously. In the 1850s only three state agricultural colleges existed in the United States.[2] The Morrill Land-Grant Act of 1862 resulted in the rapid rise of more publicly funded American state agricultural colleges, the first being the Massachusetts Agricultural College, established in 1862, followed by the Connecticut Agricultural Experiment Station, founded in 1876. Others followed, and increasingly these began to undertake various investigations – in the field and the laboratory.[3] The Hatch Act of 1887 stimulated scientific research in the United States by allowing federal funding to support experiment stations that were attached to state agricultural colleges. By the early twentieth century there were forty-eight such institutions doing agricultural research in the United States. In Canada, the Ontario Agricultural College, with its associated experimental farm, undertook experimental projects relating to agriculture in the 1870s.[4] In 1886 the Canadian government established a central experimental farm at Ottawa, where, within a short time, William Saunders produced the superior, well-known wheat named Marquis.⁵ Branch experimental farms, under the supervision of the central farm, were established later in various provinces. More Canadian agricultural colleges, normally attached to universities, were founded in the early twentieth century: by 1905 in Nova Scotia, 1906 in Manitoba, 1907 in Quebec, 1909 in Saskatchewan, 1912 in Alberta, and 1915 in British Columbia. The widespread move to put agriculture on a more 'learned' and less 'craft' basis accelerated in both countries under these conditions.⁶ Governments used the various bureaus and divisions within agricultural departments to promote ideas on how to modernize farming both scientifically and structurally. Farm activity was to be standardized, thus 'industrialized,' with the aid of advancing technology and science.⁷ The advent of genetics in the twentieth century was central to that ongoing development. Plant breeding especially attracted attention from a genetic point of view. In this chapter I outline how specific perceptions concerning heredity before 1940 shaped attitudes held by scientists towards livestock (and particularly chicken) breeding in North America.⁸ In other words, I describe the rise of agricultural genetics with a particular emphasis on its relationship to chickens.

The Beginnings of a Scientific Interest in Agricultural Breeding

The roots of North American empirical and scientific approaches to livestock and plant breeding, which had developed by the late nineteenth century, lay in eighteenth- and nineteenth-century European theoretical attitudes to speciation and natural history.⁹ As early as the seventeenth century, the philosopher Francis Bacon (1561–1626) had wondered if speciation reflected the same process which allowed breeders to use 'art' (or what we would describe as artificial selection) to alter accumulatively the characteristics of domestic animals. In 1745 the French mathematician and astronomer Pierre-Louis de Maupertuis (1698–1759) suggested that since breeders of domestic animals could change traits, why wouldn't wild animals respond to the random breeding that happened in nature?¹⁰ By the end of the eighteenth century, concerted attempts to understand heredity as an aspect of speciation led to a more widespread interest in the variation of domestic plants and animals, and ultimately to experimentation in the breeding of plants in particular. Plant breeders developed strategies that utilized inbreeding within a species and the crossing of inbred lines, or what was described by them as hybridizing. They hybridized as well by cross-breeding different species. It was the effects such breeding had on the resulting progeny that concerned them: effects that revealed secrets about the way heredity and therefore speciation worked. The German naturalist J.G. Kölreuter (1733–1806) experimented in these ways with a variety of plants. He hybridized plants to establish the existence of sexual reproduction in plants and the fixity, or constancy, of 'pure' species.[11] The British plant breeder T.A. Knight (1759–1838) studied the effects of inbreeding and the crossing of those inbred lines. (He used the garden pea, and it is most likely that Mendel knew of his conclusions.)[12] The results of such experiments revealed that certain laws, even if they were not known, governed heredity and probably speciation. In 1819 a Hungarian, Count Imre Festetics, wrote 'Hereditary Laws of Nature,' in which he argued that heredity, as evidenced from inbreeding/crossing experiments, resulted from scientific laws. He deduced four points: species inherit characteristics, they do not acquire them; traits found in past generations could reappear in much later ones; species could produce offspring with different attributes; and inbreeding must be accompanied by careful selection if one wanted to avoid the dangers of reduced vigour.[13] Even though improving crops and livestock was not the primary concern of the naturalists, some believed that experiments done to serve natural history might be utilized for agricultural improvement. Knight, for example, argued in 1799 that 'scientific' breeding of this nature could be used to help the betterment of farm plants.[14] European agricultural breeders were influenced by the views concerning heredity of naturalists like Knight as early as the late eighteenth century. Farmers in France, for example, demonstrated affinity to the philosophy of the French natural historian Georges L.L. Buffon (1707–78) when they bred livestock.[15]

By the late nineteenth century, the breeding practices of the eighteenth-century European naturalists were commonly applied to plants for agricultural improvement – that is, to increase the economically valuable traits in the stock – at state-supported institutions in North America. American agriculturalists hybridized plants (by crossing species and by inbreeding within a species and crossing inbred lines).[16] As early as the 1870s W.J. Beal crossed single inbred lines of corn at the Michigan Agricultural College.[17] W.J. Spillman first focused on the production of hay, that critical fodder crop needed to maintain horses, animals that provided the traction power of the nineteenth century. He undertook experiments at the Washington State Agricultural College in the 1890s, and by 1899 had successfully hybridized wheat. His mathematical abilities encouraged him to start quantifying results.[18]

Early Experimental Livestock and Chicken Breeding in North America

Agricultural plant breeding commanded considerably more 'scientific' attention than did farm-animal breeding in North America in this period. There was little in North American animal-breeding research undertaken at state institutions that was remotely close to plant-breeding research by the late nineteenth century. Even a casual look at animal husbandry professor William Brown's reports on livestock breeding experiments, relating to cattle and done at the Ontario Agricultural College, indicates this pattern.[19] Most academic attention paid to farm animals in both Canada and the United States focused on the large livestock species and on their feeding. The Ontario Agricultural College's experimental farm and the American state colleges devoted most of their attention, within that sphere, to cattle nutrition. Researchers at these institutions credited the practical breeders of horses, cattle, sheep, and hogs in North America with scientific expertise when it came to breeding. Their approach, based on purebred breeding and the Enlightenment traditions of Bakewell, represented what was believed to be the scientific way to improve livestock. In reality, there was little to indicate how that could be the case. Experimentation in conjunction with even a theoretical form of quantification was not part of their breeding practices.

Within the livestock-research world of the agricultural colleges and experiment stations, which focused primarily on husbandry issues, chickens played only a minor role, but they did so as early as 1877. The Ontario Agricultural College offered courses on poultry husbandry in 1877, and built a poultry house in 1879.[20] In 1880 the Maine station established a flock of Barred Rocks and a poultry house. By the late 1880s, this station was testing the effects of various feeds on chickens, as was the New York station in 1890.[21] While interest in pursuing 'scientific' poultry-breeding research was considerably less than that directed at feeding, as early as the 1880s a few breeding experiments, apparently designed to learn more about the value of craft breeder practices, took place in both the United States and Canada. The research revolved around the effects of cross-breeding, and often focused on how to find the best combinations for hybrid vigour. In 1888, for example, the head of the Poultry Husbandry Division at the Ottawa farm, A.G. Gilbert (a practical poultryman with considerable experience in breeding) crossed Brahmas with Barred Rocks for improved meat production.[22] In 1892, S. Cushman, chairman of the Poultry Division at the Rhode

Island experiment station (the first to have a separate poultry division), reported on a number of breed crosses undertaken at that institution in an effort to find the most efficient way to breed stock for meat purposes.[23] By 1900 poultry work, either teaching or research, was being done at twelve state agricultural institutions in the United States, but only two of these had a person working full time on poultry issues. One was Cornell University, where J.E. Rice (educated at that university) had begun teaching courses on poultry management as early as 1889.[24] The earliest separate poultry department at a North American agricultural college was established in 1894 at the Ontario Agricultural College.

Charles Darwin and Agricultural Breeding

It has been suggested that the close linkage of naturalist study with agriculture had broken down by the late eighteenth century, even if the experimental breeding practices of each – namely, inbreeding and hybridizing – continued to be the same.[25] It seems fair to suggest, however, that the intimacy of the connection in fact strengthened over the nineteenth century, largely because of the work of Charles Darwin, found in his *On the Origin of Species*, published in 1859, and *The Variation of Animals and Plants under Domestication*, published in 1868. The fact that Darwin concerned himself with both natural and artificial selection in the breeding of species attracted the interest of breeders. Darwin studied, for example, the hybridizing work of plant breeders.[26] He also became a fancy-pigeon breeder in 1856 and undertook to learn how pigeon breeders altered characteristics of the birds. He corresponded with pigeon-breeding experts like John M. Eaton, who wrote extensively about the art of breeding good stock.[27] It is clear, however, that Darwin himself remained confused about the actual workings of heredity.[28] Practical breeders at the time sensed this, and often argued that Darwin had not introduced innovative information on how to breed. British poultry breeders, for example, believed that Darwin merely set out many patterns of heredity which they already understood after years of experience; one being that features not seen for several generations could reappear in the next.[29] Over the second half of the nineteenth century, much biological research was devoted to explaining Darwin's evolutionary theory, a situation that led before the advent of Mendelism to the rise of myriad academic theories (such as Neo-Lamarkism) concerning speciation and the ability of a species to change.[30] Some scientists saw such work as relating to agriculture.

Darwinism and agriculture became more intimately interconnected with each other academically in both Europe and North America through the particular vision of the European plant physiologist Hugo de Vries. Trained under the German scientist Julius von Sachs, de Vries believed that biological experimentation in plant physiology should be utilized to aid agriculture, and thought so even more after becoming fascinated with Darwin's work in the 1870s (and meeting Darwin). This linkage of Darwinism and naturalist investigations involving plant physiology to agriculture laid the groundwork for a new relationship, but a much more complicated one than had existed in the late eighteenth and early nineteenth centuries, between scientific studies on evolution/natural history and research experiments designed to help farm breeding. Hugo de Vries believed that the distinction between natural and artificial selection could be overcome, and that knowledge concerning each could lead to knowledge relating to the other. An appreciation of how artificial selection worked could help agriculturalists find better methods of farming and, at the same time, help the evolutionary biologist understand the processes of natural selection. Experimentation, and experimentation involving hybridizing, was the key to both.[31] The idea that hybridization could be a tool for agricultural studies as well as for research in evolutionary biology fit with the rise of a new mechanistic approach to living matter: organisms could be controlled, altered, and manipulated – sentiments that were stronger in the United States than in Europe.[32] The American botanist Daniel T. MacDougal exemplified the American attitude. He was convinced that farm plants could be manipulated and that de Vries's ideas concerning experimentation (as a way to understand both artificial and natural selection and to achieve better farm breeding as well) was a sensible approach for agricultural production. Experimentation in evolutionary biology should be done at the agricultural experiment stations, MacDougal thought, because a critical economic result of such experimentation would be the discovery of better breeding for farm crops. The advent of Mendelism further complicated the picture.

The Birth of Genetics: Mendelism and Agricultural Breeding

After de Vries (along with several others) rediscovered Gregor Mendel's 1865 paper on the breeding of peas in 1900, he saw research involving evolutionary biology, studies on Mendelism, and efforts at better agricultural production as one and the same thing. Within that framework, he increasingly emphasized the late eighteenth-century research practice of hybridizing in breeding experiments devoted to evolutionary biology, Mendelism, and agricultural plant breeding.[33] Mendel had used the standard eighteenth and nineteenth-century inbreeding/crossing method of hybridization in his mid-nineteenth century experiments with peas. He had also concluded much of what had been observed by the earlier hybridizers. As a result, the question of exactly how innovative Mendel's work was has been debated ever since his paper was rediscovered.[34] Because he quantified the observations that he and others had made, he was able to establish laws that govern heredity. Although Mendel himself did not put the laws in the following terms (words such as gene were not in use for some time after Mendel's death, for example), our present-day interpretation of his findings is that when the gametes (or reproductive cells) form, the gene pairs separate (each unit of the pair is either recessive or dominant); and that genes are both immutable and act independently (a view that subsequently has become somewhat modified). Mendel's laws are the foundation of the science of genetics.[35]

The tighter linkage of evolutionary (designed to explain speciation) and Mendelist hybridizing experiments with agricultural breeding research by men like de Vries and MacDougal further entrenched (and provided greater credibility for) hybridization as a scientific breeding tool.[36] Mendelism gave agricultural hybridizing experiments a degree of prestige which had not been present earlier, and consequently the number of agricultural hybridization experiments at various experiment stations across the United States increased dramatically, especially in relation to corn breeding.[37] The fundamental thinking behind Mendelist hybridizing experiments, however, worked antagonistically to the approach of farm-animal breeders to the problem of heredity in a very critical way.

In hybridizing experiments, Mendelists tended to see inheritable traits as being simply recessive or dominant in character. Traits, it was argued, passed from one generation to the next on the basis of single units which were either recessive or dominant in nature. This assumption formed the basis of unit-character theory. Practical animal breeders tended to see most agriculturally valuable traits as being inherited, not in a single fashion, but rather in group fashion. Experience had taught breeders that such traits were often expressed in association with others. Cows that gave more milk often looked different, for example, from cows that were poor milkers. Chickens that laid more eggs usually had a different body shape from those that laid few eggs. Ultimately geneticists would come to recognize that the breeders' attitude was the correct one when it came to heredity characteristics relating to production in farm animals. Economically valuable traits in livestock and plants are not inherited on a single-gene basis, and it became evident later that many such traits are also not even highly inheritable. Early emphasis on unit characters by Mendelists in agricultural experimentation did virtually nothing to help identify how the economically relevant characteristics, particularly in animals, were inherited.

The Development of 'Genetic' Concepts

The British scientist William Bateson, father of genetics (he invented the word in 1906), was one of the earliest Mendelists to begin animal-breeding experiments. Bateson had studied both plant and chicken breeding before the advent of Mendelism, and shortly after 1900 he was able to show, with his poultry data, that Mendelism applied to animals as well as plants.[38] Bateson's research after 1900 relating to chicken breeding was designed to show recessive and dominant characteristics in poultry along the lines of Mendelian theory, more than it was meant to improve the birds for agricultural production. Bateson practised intensive inbreeding in order to create what he described as 'pure' lines or strains within one breed. They were 'pure' because the inbreeding (which restricted the input of outside characteristics) guaranteed that the line would breed true to certain desired observable features – for example, colour, size, or whatever else had been selected to be made uniform in that strain. This was important to Bateson, because only through crosses of lines bred for what he believed were different 'pure' unit characters could he see the expression of recessive or dominant traits.

A line that bred truly for a specific feature Bateson called 'homozygous' for it. A line that did not breed truly he called 'heterozygous' for that feature. Bateson believed that the presence of homozygosis could be explained as follows. What we would call a 'gene' (the word was invented in 1909 by the plant breeder W. Johannsen) carried two identical 'alleles' (another word that Bateson invented in 1902) for the same trait, meaning the individual had inherited the capacity to pass on only one of two possible characteristics to its progeny. Conversely, heterozygosis meant to Bateson that a gene carried an allele for a trait in a dominant fashion, and an allele for a trait in a recessive fashion, making the individual capable of passing on one or the other of two characteristics to its offspring. Bateson inbred to remove heterozygotes for particular traits from the strain.[39] It was homozygotes that revealed recessively or dominantly inherited characteristics, such as colour coat patterns. This breeding strategy was utilized by other geneticists, and generally speaking, for many of the same reasons.

Geneticists, working along the same lines as Bateson, adopted his definitions. They also introduced others to describe patterns of heredity. An important pattern that emerged in some experiments was labelled 'nicking.' Animals or plants that 'nick' well when bred to each other produce progeny which tend to carry the best qualities of each parent. Nicking can, of course, be applied to lines or families as well as individuals. Superior nicking might arise from any form of blood-related or non-blood-related mating systems. Geneticists found quite early, as had craft breeders before them, that inheritance could be either sex-linked or what geneticists defined as 'autosomal,' meaning not sex-linked. The idea of 'synthetic lines' and 'pure lines' would become increasingly important to geneticists. The phrases need to be explained and separated from each other. A synthetic line was composed of a group of individuals within a variety that bred true to the line, a line that had been established from various blended crosses of several breeds. Pure lines were different strains created within one breed, but these too bred truly to the characteristics that were wanted. Inbreeding might, or might not, go into the creation of synthetic lines or pure lines.[40] The idea of a synthetic and even a pure line was not linked by scientists to the idea of 'breed.'

Quite quickly geneticists invented new expressions to explain other theoretical concepts. In 1909 W. Johannsen proposed the word 'phenotype' to describe what a plant or animal looked like, in order to separate that perception from what he called 'genotype,' meaning the genetic make-up of the individual.[41] This idea would be of profound importance to the future development of genetics, because it emphasized that an organism should be seen as a whole, genetically speaking, and not as a set of unit characters working independently. The theoretical shift would ultimately remove barriers, set up by unit-character theory, between genetic and practical livestock breeding approaches. Geneticists came to argue that an organism's phenotype resulted from the interaction of its genotype with environmental factors, a concept that made sense to farm breeders.[42]

While Mendelists used hybridizing primarily to reveal dominant and recessive characteristics, they recognized, as had the earlier naturalists, that such crossing practices brought out other interesting results in the progeny. The hybridized progeny often held superior vigour to that of either parent line.[43] However, using the progeny to breed for the next generation led to unpredictable results. The phenomenon of increased vigour from hybridizing and the failure to maintain any uniformity over succeeding generations became an object of intense study in the United States, but not for agricultural-production reasons for a considerable length of time. Plants and animals that resulted from hybridizing were called 'hybrids,' not 'cross-breds.' The word hybrid, just as synthetic or pure line did, implied something different than the word 'breed.'

Corn Breeding and Genetics:
Hybrid Breeding for Agricultural Improvement

As early as 1908, an American geneticist, G.H. Shull, began to wonder why improved fertility and vigour resulted from the crossing of inbred parental lines of corn. He had noted the year before that such progeny could not sustain that superiority when bred with each other for the next generation. Shull knew as well from his work with corn that the greater the general homozygosis of the strain the less vigour the plant had.[44] It was clear to him that lack of overall heterozygosis, which resulted from inbreeding, meant deterioration. But Shull believed it was too simplistic to explain the superiority of the first crosses purely by the presence of heterozygosis, although clearly the return to a heterozygous state played a role in that phenomenon. He created a new word to describe this characteristic of hybridization, rather than referring to it simply as general heterozygosis.[45] In 1914 he coined the word 'heterosis.'[46] But Shull, much like his fellow Mendelists of the early twentieth century, was more concerned with proving genetic laws than he was with corn improvement. 'For Shull corn was the window, not the landscape,' as one historian has put it.[47] He wanted to understand the process of Mendelism.

Corn breeding experiments continued to be attractive to geneticists interested in the effects of both inbreeding and heterosis. One important geneticist working in the field was an American, D.F. Jones. In 1917 he reasoned that dominant/recessive characteristics of alleles played a role in heterosis. The expression of dominant alleles over recessive ones in the heterozygosis state resulted in hybrid vigour because the most dangerous and debilitating alleles for an organism always tended to be recessive. Because they were recessive, they did not express themselves unless the strain had been bred to be homozygous for that recessive. 'If, for the most part,' he wrote, 'these favorable characters are dominant over the unfavorable (if normalities are dominant over abnormalities) it is not necessary to assume complete dominance in order to have a reasonable explanation for the increased development of [the first generation] over the parents or any subsequent generation.'[48] Jones believed the stunted growth of inbred corn resulted from the presence of such abnormalities. 'There is abundant evidence to show that many abnormal characters exist in a naturally cross-pollinated species [such as corn], and that they are recessive to the normal condition.'[49] It was only inbreeding that revealed such 'lethal' recessives, a pattern geneticists labelled as inbreeding depression.

The long-standing interest in the hybridizing of plants, which dated as far back as the eighteenth century, mushroomed after the heterosis work of Shull and Jones. As early as the 1920s, regardless of the fact that the process of heterosis was poorly understood, American geneticists had begun to argue in earnest that hybridizing – that is, inbreeding combined with crossing inbred lines – for heterosis should be applied to the production of crops, most particularly corn but others as well. One of the first American scientists to believe that breeding for heterosis via hybridizing had agricultural potential was D.F. Jones, who, unlike Shull, was as interested in better corn yields as in the causes of heterosis. He realized he had to overcome the inbreeding depression from the intense inbreeding of single strains in order to utilize the process of heterosis properly, or to prove that breeding for heterosis could bring about greater productivity than simply selecting for strains that yielded more. Jones hypothesized that a double-cross hybrid method of breeding would overcome the weakness of single inbred lines, which often lost sufficient vigour that few seeds could be collected for crossbreeding. By inbreeding four strains, next crossing the four to produce two lines, and then finally crossing those two lines, he postulated, he could by the first cross restore the lines to the original fertility, and by the second cross of the two inbred-crossed lines achieve superior hybrid vigour.[50] In other words, he planned to take two inbred lines, A and B, which showed so much inbreeding depression that they produced few seeds, and cross them to produce a first-generation AB line.

Two other inbred lines, C and D, would be crossed for another hybrid line, CD, which also would yield more seeds than either C or D. Next

AB and CD would be crossed to produce a viable commercial crop that maintained hybrid vigour. In order to maintain that level of hybrid vigour over succeeding generations, seeds from the commercial crop would not be used for breeding. New commercial generations would always be regenerated by stock belonging to the parent and grandparent generations. The commercial plant was seen as a terminal product. Jones had invented the double-cross hybrid corn-breeding method, a method designed to make the crossing of inbred lines for heterosis increase production and on an ongoing basis. It would be some years, and after considerable effort and expense, however, before the method was made to work more effectively than traditional plant-breeding methods which selected for lines that reproduced truly, in spite of propaganda that suggested otherwise.[51]

Mendelism and Chicken Breeding

Mendelism led to a hugely enlarged emphasis on experimental plant breeding, but it also encouraged a new focus on animal breeding at agricultural experiment stations, and under these conditions chickens commanded particular interest. Government funding for research encouraged such a trend. The increased interest in the problem of 'scientific' farm breeding generally led to the passing in 1906 of the Adams Act, which provided money for scientific research in agriculture at experiment stations. Under this stimulus, New England quickly became a centre of poultry-breeding research. The Maine station hired the biologist Raymond Pearl under the Adams Act in 1907 for the poultry department. Rhode Island hired P.B. Hadley in 1909. Full-time biologists/geneticists joined other poultry departments shortly thereafter: for example, W.A. Lippincott in 1912 at Kansas, H.D. Goodale in 1913 at Massachusetts, and L.C. Dunn in 1919 at Connecticut.[52] Perhaps somewhat ironically, though, the new breeding experimental work done on chickens by trained biologists/geneticists after the birth of genetics focused less on increasing productivity in the birds than the limited number of experiments undertaken before 1900 had. Understanding the mechanics of evolution was often behind Mendelist chicken-breeding experiments in this period.

Biologists/geneticists who joined the poultry departments of experiment stations before 1930 hoped to elucidate the process of heredity in Mendelian terms by exploring inheritance characteristics of such features as feather colouring, shape of comb, and skeletal defects.

Examples of other characteristics studied, with Darwinism in mind, were flightlessness, crooked neck, feathering, silkiness, ragged wings, feathered shanks, multiple and double spurs, blindness, and dwarfism. Inheritance of characteristics on the basis of sex also interested early geneticists.[53] Inbreeding and crossing of inbred strains were undertaken in order to reveal the dominant/recessive nature of the traits. Goodale attempted inbreeding and cross-breeding experiments in order to assess the decline of fitness and its recovery. These studies were designed to explore both the process of heredity and the overall genetic constitution of poultry, without any consideration for the economic value for chicken breeders of the traits being studied. Scientists vaguely hoped, though, that accumulatively such work would elucidate chicken genetics and therefore how to make breeding methods help agriculture. The main breeding efforts undertaken until the late 1920s in both the United States and Canada that aimed at aiding poultry farming in the more immediate future were directed primarily at the building of good strains for distribution among farmers.[54] The resulting lines, generally speaking, came as much from traditional good craft-breeding practices as from knowledge arising from experimental Mendelism.

The work and thinking of L.C. Dunn up until 1930 serves as an example of geneticist emphasis on Mendelism in chicken-breeding experiments. Dunn undertook a number of experiments for the Connecticut research station at Storrs between 1920 and 1928 in which he investigated, for example, the inheritance of plumage colour patterns in the classic Batesonian tradition. Some of Dunn's work involved inbreeding and subsequent cross-breeding of inbred lines in poultry, but this was not done to aid farm production. Rather, it was undertaken in order to explain Darwinism via Mendelism; in this case, the process of fitness decline from inbreeding and its recovery with cross-breeding – the same process that Goodale had studied in a limited way a few years earlier. The phenomenon of fitness decline and recovery, as a process which might relate to speciation, would interest scientists working with chickens for some years to come. Dunn ran experiments as well studying the relationship of hatchability to egg weight, which he believed to be, correctly as it turned out some thirty years later, of importance to evolutionary theory. He looked into skeletal variations, the presence of lethal genes, egg-laying patterns of different poultry breeds, and the colour of the leg shank.[55] While Dunn's chief interest was Mendelism, some of the results of this work would be utilized in later quantitative poultry-breeding programs. In the 1920s, however, such information provided no aid for the farmer who bred chickens. Dunn believed as well in theoretical, rather than practical, poultry instruction in agricultural colleges. Courses, he stated, should be based on the teaching of theory/ principle, and not on the 'combined judgments of successful breeders presented as short-cuts or formulae for immediate application.'[56]

The focus behind chicken-breeding experimentation in other countries did not vary from that in the United States. The main centre abroad for research in chicken genetics between 1900 and the 1920s was Britain, and here Cambridge University was the most important. The focus there was not on agricultural production. In the 1920s and 1930s a few scientists in Poland and in Germany published papers on the genetics of the fowl,[57] and these men too hoped to explain patterns of inheritance as a general process. In 1927 Dunn toured centres of genetic research in Britain, Denmark, Norway, and Germany under a fellowship funded by the Rockefeller Foundation's International Education Board. While he found the greatest emphasis at all centres was on plant breeding, it is clear from his report that if experimental breeding was practised on poultry, it revolved around efforts at understanding the processes of Mendelism and evolution. Dunn also visited Russia. He was particularly impressed with the poultry work being done at the animal-breeding experimental centre, Anikowo, near Moscow. Under the direction of A.S. Serebrovsky, poultry served as experimental animals for studying the process of evolution. Utilizing the endless variety of chickens that could be found in southern Russia (because of the 'absence of pure breeds or pure breeding,' according to Dunn), Serebrovsky initiated the study of gene geography, namely, the study of genetic variation of a species by geographic area. Other poultry genetic investigations focused on colour inheritance (which often related to gene geography) and sex-linked characteristics. Poultry features of economic value, such as level of egg laying, were not undertaken at this Russian centre. Following Dunn's trip, both the USDA and the Animal Breeding Research Department at the University of Edinburgh in Scotland became interested in translating, in particular, *Genetics of the Domestic Fowl: Memoirs of the Anikowo Genetic Station near Moscow*, which described some of the results of experiments done at Anikowo.[58] Scientists studying chicken breeding before 1930 in the United States, Europe, and Russia pursued information that might advance knowledge of the heredity process and of the mechanics of evolution as well, but did not relate easily to the improvement of farm-producing poultry. Virtually all studies of chicken genetics between 1900 and 1930, even if their pretence suggested otherwise, did not relate to agriculture. In order to understand how genetic approaches to chicken productivity evolved, one must look outside the specific work done on poultry between 1900 and 1930.

The Rise of Agricultural Genetics and the Work of Castle

The roots of a genetics branch that would be aimed at farm production and not either pure Mendelism or the process of evolution, that is, agricultural genetics, lay in the thinking and experimental work of an American, W.E. Castle. He was one of the first scientists to address the issue of how academic genetics could (and perhaps more importantly should) improve the breeding of livestock. While Castle was primarily interested in Darwinism and the process of evolution, the potential relationship of genetics to agriculture also concerned him, and this focus would ultimately affect his student, Sewell Wright. 'The existence of civilized man,' Castle wrote in 1911, 'rests ultimately on his ability to produce from the earth in sufficient abundance cultivated plants and domesticated animals. A knowledge of how to produce useful animals and plants is therefore of prime importance.' He acknowledged that master breeders in the past had been able to create all modern improved breeds of livestock (lines that bred truly or were homozygous) without the input of genetics, and that they had done so by using some form of inbreeding.[59] Castle admitted that breeders knew certain sanguineous lines did not lack vigour, and that careful selection in the beginning could help sort out degenerative lines. He also realized that farmers regularly crossed breeds for greater productivity and knew such crosses should only be used for market purposes. They understood, in other words, that the crossed progeny should not be used for breeding purposes because such matings would not result in progeny that presented either uniformity or increased vigour. 'The production of [animal breeds] and their use in crosses is both scientifically correct and commercially remunerative,' he noted.[60] For Castle, the problem was that breeders did not know why any of the results occurred. 'The present is an age of science; we are not satisfied with rule-of-thumb methods, we want to know the why as well as the how of our practical operations,' he argued.[61]

Castle studied the historic effects of inbreeding on domestic livestock. He believed that an understanding of the process would explain why the breeders had been able to improve farm animals. But like de Vries, Castle believed the study of artificial selection could help scientists understand the way natural selection worked. How to utilize inbreeding experimentally, it seemed to Castle, was the fundamental issue at hand in studying the mechanics of evolution, Mendelism, and ultimately improved agricultural production. He undertook breeding experiments with guinea pigs that relied on inbreeding to generate both self-perpetuating lines and hybridized lines. Castle's true-breeding-line work led to the development of specialized laboratory mice. His student, C.C. Little, bred mice populations that were designed for use in medical and human genetic experiments; and under these conditions true breeding lines were essential.[62] The potential of inbreeding and the crossing of such lines within one species for Mendelist purposes, however, also attracted Castle's attention.[63] His inbreeding and line-crossing research influenced Sewell Wright, whose theoretical genetics ultimately lay the groundwork for agricultural genetics applicable to livestock.

The Rise of Agricultural Genetics and the Work of Wright

Sewell Wright is best known as an evolutionary geneticist, and his primary interest throughout his life remained the mechanics of evolution rather than livestock breeding. But in his early professional years Wright studied livestock production (especially the historic breeding of Shorthorn cattle)[64] and also worked for the US Bureau of Animal Industry. He wrote articles in the livestock journal the *Breeder's Gazette*, although his language was so specialized that breeders would have found little of the information helpful. Wright looked at the effects that assortive mating based on phenotype (choosing an animal because of its looks, with no regard for its genetic background), as opposed to inbreeding, had on the level of uniformity (or homozygosis) found in subsequent generations of farm animals. He proved that similar looks did not necessarily imply shared genes, and that phenotype could not be relied on as an indicator of genotype. While Wright failed to attract much notice from practical breeders, his theoretical work proved to be critical to the evolution of a form of genetics that would draw the attention of the agricultural world.[65]

One of the most important things that Wright did for agricultural genetics, and subsequently for future chicken breeding, was to quantify the effects of various inbreeding strategies.[66] His work in effect quantified earlier breeder theory, and provided a more complicated way of controlling the level of inbreeding, something that Sebright had recognized as important as early as 1809, and Bakewell even earlier. Perhaps one reason why future theoretical animal breeders, that is, agricultural geneticists, came by the 1940s to see Wright as the founder of genetics for farm-animal breeding was the fact that his path-coefficient calculations for inbreeding reflected, in a way that the work of other geneticists did not, the refined attitude to the interbreeding of blood-related individuals that historically all the great breeders of the past had adhered to.

Using statistics obtained from Castle's inbreeding experiments on guinea pigs along with his own experimental results, between 1915 and 1922 Wright devised a way of calculating the level of shared genes that would result from different inbreeding systems – brother to sister, first cousins, double first cousins, half-brother to half-sister, and so on.[67] The path coefficient allowed a scientist to predict how much homozygosis or return to the genetics of the original cross would result from matings after that first-generation cross. Wright had established an accurate quantitative way to measure the level of homozygosis resulting from any type of inbreeding. Originally he had believed that careful selection of stock should be practised along with blood-related breeding in order to fix superior type, but later he came to think that the mating of animals related to each other to some degree was more important than selection, even though he accepted the theory of inbreeding depression, or the rise of lethal recessives in inbred lines.

Heterosis also interested Wright. He was well aware of Shull's work with inbred corn and knew from his own experiments that crossing inbred lines often led to progeny superior to either parent.[68] He wrote: 'By starting a large number of inbred lines, important hereditary differences ... are brought clearly to light and fixed. Crosses among these lines ought to give a full recovery of whatever vigour [had] been lost by inbreeding, and particular crosses may be safely expected to show a combination of desired characters distinctly superior to the original stock.'[69] Wright believed his path coefficient might work well for the production of synthetic lines designed for crossing in order to achieve heterosis in the resulting hybrid stock. He could quantify the level of inbreeding and thereby reduce its intensity, thus avoiding some of the dangers it could incur. The path coefficient in the end would make it easier to create vigorous lines resulting from matings of related animals that could be used to cross for heterosis. (The coefficient could equally well be used to produce superior pure lines designed to breed on truly.) Research geneticists began to explore how hybridizing could work within the controlled framework of Wright's path-coefficient theory.[70] The theory did not, however, reach the breeders with any clarity for some years.[71]

Wright's other major contribution to agricultural genetics and ultimately chicken-breeding methodology resulted from his work in population genetics.[72] It was the union of Wright's path coefficient with population genetics (a new arm of genetics which Wright also initiated, but along with R.A. Fisher and J.B.S. Haldane in Britain) which began a pattern that would ultimately bring livestock breeders, first in the United States and subsequently in other countries, into the world of genetics. Population genetics, that is, the assessment of inheritance patterns on the basis of groups not individuals, played an important role in a large movement outside livestock breeding which occurred after 1930 and was known as the evolutionary synthesis. The foundations of population genetics had been laid down before the advent of Mendelism, and evolved from mathematic systems designed to assess hereditary patterns. Labelled biometry and developed by Karl Pearson in Britain, the approach had competed since 1900 with Mendelism for attention, especially in relation to the process of evolution. Scientists interested in genetics, various other aspects of biology, or evolution would be able to reconcile Mendelism to Darwinism through population genetic theory.[73] The schism between genetic and practical animal breeding in North America, which dated from Mendel, remained in place longer, until statistical systems could be developed for agricultural breeding systems. Statistical systems useful for animal breeding arose from both population and what is known as quantitative genetics.

The Rise of Agricultural Genetics and the Work of Lush

An American, Jay L. Lush, utilized the theories of Wright to develop what ultimately became agricultural genetics applied to the breeding of farm stock. Lush combined theoretical population genetics with Wright's path coefficient in order establish safe ways to inbreed large groups of animals. But equally significant for the future of livestock breeding, he began to change his approach within that framework. Lush started to emphasize quantitative genetics over population genetics.

In fact, until geneticists could focus on animal heredity as a quantitative issue, they could do little to affect the process of breeding livestock. Quantitative genetics assesses the degree to which animals of a species inherit a certain characteristic which all members of that species inherit to some degree, and measures individuals against the average of a population. Farm animals within each species inherit characteristics that are marketable in varying degrees. Farmers wanted, for example, to increase milk yields in a cow, and since all cows provided some volume of milk, it was a question of the volume of milk given by different cows; therefore, a quantitative problem. Similarly, all hens laid eggs. It was a quantitative question as to how many – hens that laid more eggs were preferred over those that laid fewer.

There were other distinguishing features that divided quantitative genetics from population genetics. Quantitative characteristics normally result from the combination of many genes, and therefore quantitative genetics looks at the effects of inheritance involving many loci. Population genetics, while it addresses organisms in groups and does not concentrate on individuals, focuses on qualitative traits – traits that often describe speciation – and tends to look at inheritance patterns at one locus, or point, where the alleles join. Population genetics is the Mendelian base of the study of evolution, but quantitative genetics also rests on a Mendelian base. The first attempts to treat Mendelism quantitatively were initiated as early as 1909, but for some reason attracted virtually no attention from geneticists. The earliest paper that looked at population genetics within Mendelian considerations was published in 1916, and was widely read by scientists.[74] The real theoretical foundations of both quantitative and population genetics, however, developed in the 1930s from the work of Wright, Fisher, and Haldane.[75] The new orientation to quantitative genetics and away from population genetics would help scientists to see farm-animal breeding outside the question of evolution. This proved to be a very important shift in outlook, because it allowed geneticists to detach animal-breeding experiments from studies in evolution. Research devoted to animal breeding would no longer be perceived as being one and the same thing as research into speciation.

Lush came from a farm background and was trained at Kansas State Agricultural College in animal husbandry. Through a professor at the college, who taught animal breeding and was a friend of Wright, Lush began corresponding with Wright as early as 1918. Between 1918 and 1922, while Wright developed his path-coefficient theory of inbreeding and wrote about systems of mating, Lush kept up with the literature as it appeared, and quickly saw that much of it was applicable to livestock-breeding strategies on farms. He took courses on evolution as well from Wright, and found Wright's work on inbreeding in historic Shorthorn cattle of great interest.[76] At Iowa State University, Lush assessed inbreeding levels practised on other livestock species. He synthesized the theories of Wright with statistics and theory arising from embryonic population and quantitative genetics, and what he could learn from Mendelist geneticists at the college into a comprehensive animal-breeding theory that could be utilized on the farm.[77] At Iowa State he gathered together graduate and post-doctoral students who would learn his theories and take them literally all around the world.[78] Lush always felt indebted to Wright's work, even though Lush himself actually created a more useable and practically oriented set of theories, embodied in quantitative genetics, designed to improve farm animals.[79]

In the 1930s Lush wrote extensively about animal-breeding methodology and the historical background to livestock breeding within that context. He argued, fundamentally, that the methods of selection that he advocated had been known by breeders for centuries. The main innovation, for Lush, was the fact that Mendelism and its extension to large population bases via quantitative genetics explained why certain results occurred when different selection methods were used and, perhaps more important, suggested ways of predicting what might occur when a certain method was followed. It would be possible, Lush believed, to learn which method of selection worked the best in a breeding program. Predictability, Lush believed, was important because it saved time and expense. While time and expense incurred by experimentation with varying methods were beyond the reach of farmers, experiment stations should undertake them.[80] Lush fully subscribed to the idea that government and science should work together to promote better agricultural production. Improvement, at this stage, was unlikely to be generated by farmers themselves.

Lush's experience in the field of practical breeding, combined with his training in developing genetics – in particular, Mendelism, statistics, and both embryonic-quantitative and population genetics – gave him a historical perspective on the problem of how genetics could and did interface with craft in animal breeding. He moved effortlessly between the two, often conflicting, worlds. Lush noted that the obsession with race fixity that many European writers of the mid-nineteenth century demonstrated was unlikely to have commanded much interest from contemporary breeders.[81] Experience had shown them that they could alter features in animals. Race plasticity dovetailed with the idea that it was possible to alter characteristics in domestic animals. Lush also explained that early research in Mendelism contributed three significant facts to early-twentieth-century purebred livestock breeders. First, identical pedigrees did not mean identical heredity. Second, there was a distinction between genetically caused and environmentally caused variation. And third, Mendelism explained why, over generations, individual variation within a population did not appreciably change.[82]

Lush was as concerned with the process of heterosis as with the mechanics of inheritance via quantitative genetics, population genetics, or Wright's path coefficient. (By the late 1940s new computation methods for understanding inbreeding levels overcame the unwieldiness of Wright's system when it was used in ever larger breeding populations.)[83] Chicken breeding in particular and the effects of heterosis on the birds interested Lush. He experimented with poultry breeding at Ames in 1945, attempting to produce chicks along the lines of hybrid corn breeding, namely, by inbreeding various lines and crossing for heterosis.[84] He was not alone in this poultry breeding work by that time. A number of geneticists (D.C. Warren, a poultry geneticist at the Kansas State University, is an example) were highly focused on applying the hybrid corn-breeding method to chickens by that time.

It was, quite simply, possible to experiment quantitatively with chickens via inbreeding and hybridizing in a way that was not true of the large livestock. In contrast to large livestock, fowl were cheap individually speaking, and poultry reproduced relatively quickly in sufficient numbers to warrant quantification of results. As Lush explained some years later at a fact-finding conference of the Institute of American Poultry Industries: 'With the larger animals, the volume of business, the time demands, the value of the individuals, is so great as to make one hesitate a lot toward [looking for large numbers of strains or numbers within a family]. You in poultry, I think, are pioneering in that direction as far as the animal field goes.'[85] Lush recognized that the cost factor, relating to the time needed to apply breeding experiments and quantify results for larger farm animals, made such work daunting for either small breeders or government.

Genetics and Livestock Breeding at Experimental Stations before 1940

While chicken breeding received increasing attention from geneticists working at agricultural colleges and experiment stations, livestock breeding experiments undertaken by government institutions, especially those relating to the large farm species and based on Mendelian strategies, were not common in the first half of the twentieth century. A survey done by the USDA, for example, revealed in 1936 that of eighteen experiment stations, only five did any breeding experiments with beef cattle. Of forty-eight experiment stations surveyed, only nine undertook any swine breeding experiments, and most of these did not relate to better economic traits. In fact, before 1920 only the Delaware station had given any attention to this problem. Inbreeding in swine seemed to bring out lethal disease factors. Very few experiment stations took on sheep-breeding experiments, and when they did so most did not rely on inbreeding systems, a factor which seemed to the USDA authorities to make the experiments of little substance.[86] After all, inbreeding and hybridizing was *the* way to breed experimentally and therefore scientifically. The tendency to view hybrid corn breeding as the model for Mendelism, and consequently as the scientific approach to breeding, made many argue that any other strategies used for breeding did not reflect Mendelian principles and therefore were innately lacking a scientific basis.[87] Genetics and genetic experiments did little to alter the way farm animals were bred by the late 1930s because it was virtually impossible to use the hybrid-corn-breeding method on them. The American *Yearbook of Agriculture* for 1936 pointed out that there had been up to that time virtually no application of genetics to animal breeding.[88]

Some agricultural experts believed in a vague way that better results would eventually emerge from genetic research,[89] but others argued that would never happen via the hybrid-corn-breeding method. It could not be applied to large animals, they believed. Experiments resting on inbreeding and hybridizing were too costly with the slow reproductive rate of the large animals, and such breeding as might take place at the stations reflected too many environmental variables. Many suspected also that the culture of livestock breeding was too entrenched in an ethos of another era to undergo the radical changes in outlook that adherence to the hybrid-corn-breeding method entailed. Scientists who continued to insist that inbreeding was essential for improvement (it was an integral part of the hybrid-corn method) also found their work hampered by the fact that breeders strongly resisted inbreeding. American livestock breeders had moved away from that methodology over the nineteenth century.

It did not help that historically farmers and livestock breeders had always been reluctant to adopt any method of farming that was deemed 'scientific' across both the United States and Canada. The failure of much of nineteenth- and early-twentieth-century science to improve upon agricultural practices that had been learned through experience tended to make farmers, with good reason, impervious to innovative ideas.[90] The fact that geneticists could not offer any concrete information with respect to the larger animals did not encourage farmers to look to any potential knowledge emanating from science in the future for the breeding of large livestock.

The poultry situation was different. Chicken breeding had become common at experiment stations after the rise of Mendelism, and the new approaches that emanated out of the theories put forward by Lush escalated that trend. Agricultural genetics (with its emphasis on population/quantitative genetics) quickly infiltrated chicken-breeding methodology – from both a theoretical and experimental point of view – at the agricultural experiment stations. Genetics as applied to any agricultural research ceased, generally speaking, to be based on simple Mendelism by the 1930s. Within this environment, scientists studying chickens no longer took an interest in collecting data on the mode of inheritance of distinct traits.[91] Complicated inbreeding and outcrossing programs, designed to utilize hybrid vigour, increasingly dominated research experiments at the colleges and experiment stations. Hybrid corn breeding had become their model. Many scientists were discouraged by the results. Hybrid performance of chickens might have been better than that of the inbreds in the experiments, but it did not match that of existing superior, non-inbred stock. The expensive and time consuming experiments often ended up with stock that, after returning to normal vigour, reached a level of productivity which had existed before any breeding system had been applied to them. Many of the inbred crosses were done within a strain (thereby weakening the potential for heterosis), and in most cases no attention was paid to the performance of the birds used to create the inbred lines.[92] But, while results from inbreeding/hybridizing experiments were still discouraging for geneticists at research stations in the early 1930s, hope was on the horizon.[93] This shift in scientific focus, however, did not bring a greater affinity between the geneticist and the untrained breeder. The divide between the two would only continue to widen as expensive quantitative experiments got under way. New developments, which had nothing to do with genetics per se, would lead to the fulfilment of that hope, as will be evident later in this book. The effects of evolving genetic theory and practice on craft breeding and culture between 1900 and the 1940s are the subject of the next chapter.

Chapter Four

Breeding for Eggs in North America: Conflict between Science and Craft

Mechanization of agriculture and standardization of production, both of which gathered momentum early in the twentieth century, were keys to the growth of all aspects of farming in the United States and Canada.[1] As the chicken industry entered the twentieth century, technology lent urgency to the campaign to standardize and improve production because mechanization alone allowed for unprecedented expansion. Technology which encouraged production on a seasonally even basis laid the foundations for large-scale commercialization of the egg trade. More efficient transportation via railways played a role in making eggs available year round, but it was the invention of cold storage just before the twentieth century that was most critical to that development. By 1890 refrigerated warehouses, cooperatively or commercially owned, existed in the United States. These provided cold storage for perishable produce for a considerable length of time. The Canadian government offered aid in the 1890s for the building of refrigerated warehouses at various locations. Because cold storage meant eggs were available on a more even basis throughout the year in both North American countries, it greatly expanded the marketability of the product, domestically as well as internationally. Consumption rates rose accordingly in both countries. In Canada, for example, egg consumption in cities and farms increased from twelve dozen per capita in 1891 to fourteen dozen per capita in 1901 to seventeen dozen per capita in 1911.[2] Cold-storage and marketing systems functioned in a similar fashion in Canada and the United States. Cold-storage operators began to fill their plants in March and, at an increasing rate, over April and May, then reduced their intake in June and July. Storage was full by August and began to be sold. Sales rose over September and October, and reached their height in November and December. Usually supply lasted until the following March, when the intake began again. Eggs normally sat in storage for up to six months. April eggs were considered the best, and they were generally the most numerous as well. The quality of eggs stored, however, was often questionable. It was estimated in Canada in 1914 that not over 33 per cent of eggs received in large markets were excellent, the rest being of varying degrees of poorness. Some dealers in eggs found that as much as 10 to 12 per cent of their receipts were unfit as food. In 1913, the USDA announced that at least 17 per cent of American eggs that came on the market were unfit to eat, and in 1914 poultry expert M.A. Jull reported much the same thing.

Part of the problem of poor eggs was that there was no incentive for the egg producer to sell only those of good quality. Country stores traded in eggs on a barter or cash basis, and their interest lay in bringing in customers, not in selling high-class eggs. Under these conditions, farmers received the same amount for good and poor eggs. The complicated marketing structure through country store, passing trader, and local buyer to poultry and egg packer did nothing to encourage either good quality or uniformity of eggs. Grading was done by the packer who shipped the eggs forward to the final buyer.[3] The marketing of eggs in this fashion, while it might not have resulted in uniformly high-class eggs, did, however, play a critical role in the way the overall farm family economy worked. Barter in eggs brought necessities into the home when no money was available, and also generated exchange in goods that stimulated local trade in a variety of ways.[4]

The potential for year-round sales of eggs in urban centres that cold storage encouraged not only resulted in enlarged markets, but also seemed to play a role in fluctuating trade patterns within North America and abroad. From the American Civil War in the 1860s until 1900 Canada exported many eggs to the United States, but subsequently the trends of intercontinental trade shifted. By 1902 Canada was importing from the United States and exporting to Britain. But import rates for American eggs into Canada exceeded those for export to Britain and the United States combined.[5] This situation might have seemed serious to Canadian poultrymen, but it did not impact their American counterparts in a similar fashion. The United States export rates over the period fluctuated with American under/over domestic supply, and always represented only a small proportion of poultry-industry output. American import rates were also minimal. Imports never exceeded more than 2 per cent of domestic production during these years. The American export/import trade in poultry products was less significant to the economy of the American chicken industry than it was to that in Canada.

The First World War stimulated the expansion of the egg industry that cold storage had encouraged. It opened up an increased Canadian market for eggs in Britain, for example. The number of Canadian chickens rose from its pre-war 1911 level of 30 million to 37 million by 1919.[6] Russia, who had supplied 50 per cent of the eggs imported into Britain, had dropped out of the international market by 1919. This fact provided further opportunities for countries selling to Britain after the war.[7] A poultry boom took place during the war in the United States as well. Chicken numbers rose from 280 million on farms before the war to 359 million by 1920.[8] The American export/import market for poultry products continued, however, to be less significant to the poultry industry in the United States than it was for the Canadian industry. The growth of the American industry was mainly based on the domestic economy until after the Second World War.[9]

The advent of genetics coincided with the establishment of cold-storage plants, and it would not take long for geneticists to argue that their experimental work with chickens, even if its underlying focus was on the study of evolution, would ultimately transplant the methodology of practical breeders. Technology and expanded trade activity in eggs provided a powerful raison d'être for scientists and poultry experts working at government-supported agricultural institutions, and helped to convince them that soon they, not practical breeders, would be the experts in breeding. As early as 1911 the scientist W. Johannsen discounted even the historical input of practical breeders to the problem. In fact, Johannsen argued that Darwin himself had overvalued the importance of breeders to questions of genetics.[10] While it might have seemed that, before 1915 and the chromosome work of T.H. Morgan, livestock breeders commanded as much respect as geneticists in the matter of heredity knowledge (they were equally important to the American Breeders' Association), a cleavage between the two groups was apparent.[11]

The Use of Genetics in Commercial Chicken Breeding for Eggs: The Work of Pearl

One early geneticist in particular, Raymond Pearl, did more than give lip service to agricultural interests when it came to poultry breeding. He focused on how genetics could work with the farm breeding of chickens for the egg industry. Between 1907 and 1916, while working at the

Maine Agricultural Experiment Station, he studied the egg productivity of hens. A device known as the trapnest (inferior versions of it existed as early as 1869) aided in his research by allowing him to identify heavy-laying birds in a flock. The trapnest worked by locking a hen into a nest when she entered it to lay an egg. The breeder/producer could thus link her with the laid egg. Inbreeding and crossing for increased production via heterosis was not central to Pearl's work. He wanted instead to identify the way egg-laying inheritance worked. A biometrician who subsequently became interested in Mendelism as a result of his chicken-breeding work, Pearl believed that statistical analysis of trapnesting results could explain to poultry keepers why certain selection strategies worked better than others. He tended to see the problem of inheritance in population, not individual, terms. He therefore encouraged poultry breeders to look at inheritance as a flock problem, and also to select on the basis of what bird populations could produce, not on what their ancestry had been. He concluded as well from his biometric studies of trapnesting records kept at the station that mass selection on the basis of either individual worth or ancestry – namely, choosing high-producing hens and the sons of high-producing hens to breed for the next generation of layers – did not work. This selection method had not increased the egg-laying averages of the flock at the Maine station over a ten-year period. He believed too from trapnesting records that the progeny test was the best means of selection.

Pearl started a new set of experiments by crossing a known egg-producing breed with a non-egg-producing one in order to learn more about the process of fecundity and inheritability. Interesting results arose from the breeding of the Barred Plymouth Rock (an egg-producer) with the Cornish Game (a non-egg-producer). Regardless of the females used in breeding, pullets that came from males of the high egg-laying breed (Rocks) consistently outperformed those sired by males of the poor egg-laying breed (Cornish Game). Furthermore, the female progeny of the crosses produced in patterns consistent with their male heritage. It appeared that, while egg-laying capacity was inheritable, the female played a limited role in high egg-laying levels. With this thought in mind, Pearl next dissected a number of hens in order to see if the low egg-laying hens appeared to be less capable physically of potential production than the high egg-laying individuals. He found that physiologically the poor egg layers were anatomically identical to the superior egg layers, suggesting that the poor layers were equipped to be as prolific as the superior laying hens. The dissecting results, therefore, tended to support his view that males triggered levels of egg-laying in females. He consequently believed that more emphasis on selection practices (and via the progeny test) based on males might lead to better egg-laying flocks. By 1911 his impressions concerning fecundity and male transmission of the trait had convinced a number of agricultural experts that breeding efforts aimed at egg production in a flock should concentrate more on the breeding of cocks, rather than hens.[12] Some poultry experts in Canada advised that closer attention be given to breeding males.[13]

But Pearl's lack of clarity on certain underlying issues led to considerable confusion in the minds of others who read about his work. Pearl did not, for example, make it clear whether he had totally rejected the input of females to their daughters. Some experts continued to argue that the female was, in fact, as important as the male for egg production.[14] Pearl did not explain adequately, either, how the process of male inheritance of fecundity worked. Equally critical to ensuing misunderstandings between the scientist and the general poultry public was the fact that many failed to realize that Pearl's conclusions should be seen within the framework of the progeny test, as opposed to mass selection on the basis of individual worth or ancestry. Consequently, poultry experts often concluded that, because good breeders had always appreciated the value of the male in egg production (via his female ancestry), Pearl's theories offered nothing new. R.R. Slocum, for example, who worked for the Bureau of Animal Industry and who published a book on the breeding and mating of chickens in 1920, stated that Pearl's advice on how to increase egg production through an emphasis on males matched what any good breeder would already be doing. Breeders knew sons of good layers were important for egg production, Slocum pointed out.[15]

The pervasive ethos of purebred breeding, which had come to permeate most improved breeding of livestock over the last one hundred years and emphasized ancestry breeding, played a role in making it difficult for men like Slocum to understand Pearl's emphasis on the progeny test. Purebred breeding principles had deviated from that basic Bakewellian approach increasingly over the nineteenth century, and tended at the same time to promote selection on the basis of the individual in relation to its ancestry.

A Craft Breeder's Response to Pearl's Theories

Chicken breeders were aware of Pearl's views, and even more so by 1913 after he published in the *Reliable Poultry Journal* and spoke at the annual meeting of the American Poultry Association.[16] When Pearl's comments in bulletin 305 of the Maine Experiment Station are compared with a particularly prominent breeder's reactions to them, the nature of the difficulties and differences between the scientist and breeders become evident. Pearl stated in bulletin 305 that 'selection to the breeder means really a system of breeding. Like produces like, and breed the best to the best; these epitomize the selection doctrine of breeding. It is the simplest system conceivable. But its success as a system depends upon the existence of an equal simplicity of the phenomena of inheritance.' If, for example, a breeder mates an individual that is larger than average to another individual larger than average and always gets offspring larger than average, then breeding 'the best to the best' would, as Pearl put it, 'offer a royal road to riches.' But if, he continued,

> a character is not inherited in accordance with this beautifully and childishly simple scheme, but instead inherited in accordance with an absolutely different plan, which is of such a nature that the application of the simple selection system of breeding could not possibly have any direct effect, it would seem idle to continue to insist that the prolonged application of that system is bound to result in improvement ... It seems to me that it must be recognized frankly [that a selection method] must be based and operated on the gametic condition and behavior of the character in which improvement is sought ... Continued mass selection of [asexual] variations as a system of breeding, in contrast to an intelligent plan based on a knowledge of the gametic basis of a character and how it is inherited seems to me to be [a mistake].[17]

H.H. Stoddard, a man who had been concerned with breeding for better egg production for over forty years and who had written extensively on the subject in various poultry journals, responded to Pearl's views, thereby explaining them to other breeders. In his articles for the *American Poultry Journal*, Stoddard challenged Pearl's conclusions, questioned the innovativeness of his suggestions, and took issue with some of the inflammatory language found in the bulletin. Stoddard stated:

> The fact is the bulletin is wrong. High fecundity and low, too, may descend from either sex to either sex or it may not descend at all directly from either to either. There will sometimes [be] great irregularity, and scattering every

which way, and reversion to remote ancestor types, especially if there has been a cross of strains considerably diverse. Selection for the purpose of breeding from the best to get the best, even if it is 'childishly simple,' will continue to be the only way to fix characteristics, and among many misses there will be some hits. Breed 'the best to the best' and though you may find that some of the progeny may not be as good as the average of their parents, yet some may be as good and some decidedly better.

Stoddard pointed out that this breeding method – namely, mass selection on the basis of individual worth and ancestry – had brought the original egg production of wild fowl from six to eight eggs a year to at least and often more than fifty in domestic chickens.[18] He argued also that fecundity could be inherited from hens by their female progeny. 'I do not deny the influence of the male bird in helping to build up a strain of great laying. Neither do I know of anyone who would … What I do deny is that dams have no finger in the pie of hereditary fecundity. They have a great deal to say about it.'[19] Stoddard assumed that Pearl must mean, even if he did not articulate it, that the selection of males proceed on the basis of their mothers. 'What Dr. Pearl really teaches us is that fecundity is transmitted equally by both sexes, and by alteration,' Stoddard argued, 'but in his summary (misleading because incomplete) has laid a trap for his readers; for, although it tells what the daughters inherit, and from which parent, it is silent as to what the sons inherit and from whence.'[20] Pearl actually intended that males be chosen via the progeny test, namely, on the basis of their daughters.

Stoddard concluded that Pearl had needlessly confused the existing situation and at the same time had offered no new useful suggestions. For Stoddard, Pearl advocated what breeders had always done, namely, practise mass selection on the basis of individual worth and ancestry for both males and females. Under 'Pearlite' theory, as Stoddard put it, a heavy layer would be mated to a cock whose dam was a heavy layer. Isn't that the same as the best to the best, he queried, that is, both males and females from families in which the female members were good egg layers? It is just that Pearl called it something different. Stoddard summarized his impressions concerning Pearl's approach to breeding in the following words: 'If the "childishly simple scheme" or "breeding from the best to the best," which "is the simplest system conceivable," was so totally and disgustingly fruitless in the past, will the identical practice result differently because of masquerading under a new name?'[21] Stoddard hit on a main point when it came to comparing craft with genetic breeding: the same breeding method could be utilized by both groups but for different reasons.[22]

While Stoddard's views make it evident that experienced breeders did not necessarily grasp all the implications behind Pearl's thinking, they also show that breeders could identify underlying problems in the scientist's approach to heredity in chickens. By focusing on male input to egg laying, Pearl lent his support to the incorrect idea that the heritability of egg laying worked in a relatively simple way, an idea that Stoddard and others would never have accepted. In fact, egg-laying inheritance is genetically complex, multi-trait-oriented, and involves low inheritability, but via both sexes. And because it is difficult to estimate genetic improvement for egg laying from any particular breeding program, progeny testing should be done in conjunction with other methods of selection.[23] Pearl's work, however, proved to be valuable for a number of other reasons. First, he initiated, more than anyone else before him, what would be an ongoing and often controversial trend: namely, the attempt to make genetic knowledge work in conjunction with practice. Castle had been more likely to theorize on that possibility than to cross the bridge. Second, Pearl's work indicated that progeny testing worked better than a reliance solely on mass selection, a fact that unfortunately went unrecognized for some time, possibly because mass selection still offered room for improvement.[24] Pearl also failed to undermine breeder faith in ancestry breeding and individual worth.

Stoddard not only assessed Pearl's chicken-breeding work, but also watched the general rise of Mendelism with some interest. He saw the science as being too embryonic in its development to be useful at the current time in agricultural breeding. In fact, Mendelism was nothing more than practical breeding under a different guise, Stoddard thought. 'Mendelism, or the new genetics, or whatever it may be called,' Stoddard noted, 'offers at its present stage not new practical instructions for mating and breeding either the lower animals or humans. The professors who say that the old rule of "breeding the best to the best," is no good; turn right around and prescribe methods that amount to the same thing.'[25] But Stoddard believed that Mendelism ultimately would offer practical breeders aid. He concluded: 'The whole problem offered by Mendel's discovery, one of the most important as well as wonderful, in the annals of science, is such a complicated one that it will take generations to solve it, and at present the breeders of domestic animals ... can derive little benefit or none at all from all that Mendelism can offer – in its present stage of development.'[26]

Other breeders found Pearl's work even more confusing than had Stoddard. J.B. Morman, for example, concluded that Pearl's work implied that egg laying was not even an inheritable trait. Since experiments done in 1912 by Morman himself had convinced him that egg laying was an inheritable characteristic, he saw Pearlism as a dangerous trend, and one which had already convinced the famous biometrician Karl Pearson to 'relegate the problem of inheritance of egg-laying power in fowls to oblivion.'[27]

The Poultry Press and Attitudes to Genetics

By the 1920s the poultry press had begun to assess, but in an unclear way, how Mendelian studies had proved that inbreeding (but known as line breeding) should be practised. Sometimes the editor of the *American Poultry Journal* praised genetics for its ability to shed light on this most important breeding issue. 'A more general understanding of the benefits and limitations of linebreeding has ... assisted the breeders ... Scientific investigation has shown that linebreeding affords the most certain means of intensifying and mixing desirable qualities in a strain, and wide-awake breeders have been quick to apply this principle ... Many poultrymen are still afraid to take up linebreeding because it is a form of inbreeding and they have a wholesome fear of the latter practice,' the *American Poultry Journal* told its readers in 1921. Proper line breeding, however, required careful selection, the journal added, and the ability to select came from experience, not science.[28] Fundamentally, the journal argued that, while genetics might be able to explain the value of inbreeding, it could not tell how to select. At other times, the *American Poultry Journal* was inclined to deny that science had even contributed to an understanding of inbreeding. For example, when a breeder reported in 1922 on forty years of controlled inbreeding (he had only introduced a male twice in forty years), the editor of the *Journal* noted the wisdom that experience brought, saying, 'It is one thing to experiment with rats in a laboratory, or try to transfer to chickens an experiment with seed corn, but it would take forty years of practical work on the part of a skilful breeder to produce a testimony on inbreeding equal to the above.'[29]

Articles in the *American Poultry Journal* on how Mendelism explained inbreeding did not elucidate the results of genetic research on the subject very well, but they did heighten the confusion as to what genetics could teach breeders and how the 'genetic' differed from the 'practical' way to breed. Within that framework, the press did not appear to recognize Mendelism's very real contributions to craft breeding. Mendelism had, for example, made it clear why certain selection principles should be abandoned. For instance, it was evident from early Mendelism that remote ancestors were less important in a pedigree than a sire or dam's brothers and sisters. Mendelism also reintroduced the importance of the progeny test, but indicated that all progeny, not just the good ones, should be taken into account when judging the worth of a potential breeding animal. But while Mendelism revealed why inbreeding led to certain results, it did not address what to select for in an inbreeding program, or how inbreeding related to the inheritance of quantitative characteristics. Because science failed to address these last two factors and because they were the more critical ones for agricultural breeding programs, breeders found it hard to see what Mendelism did in fact offer them.

Geneticists further aggravated the situation by implying that certain breeding principles, long understood by good breeders, were 'scientific,' thereby adding to the difficulty of understanding what was new in genetic breeding theory. (Scientists openly acknowledged their debt to craft breeders, however, when it came to information on husbandry, feeding, and housing issues – all of which were essential to the success of experimental breeding programs.)[30] For example, the idea that the phenomenon of hybrid vigour was an innovative discovery of research scientists (a point of view that some geneticists seemed to adhere to) was problematic for experienced poultry and livestock breeders who were familiar with Bakewellian principles and Sebright's theories. Scientists had begun to breed for farm improvement by utilizing hybrid vigour, and they saw heterosis as *the* way to 'scientifically' breed superior progeny. The geneticist D.C. Warren, who graduated with a PhD from Columbia in 1923, and served as poultry geneticist for twenty-five years at Kansas State University before becoming a geneticist consultant for a breeding company, discussed his experiments in the poultry press, describing hybrid vigour in relation to cross-breeding, or inbreeding and crossing of lines, but hardly in a way that would be new for readers. 'Through close breeding or inbreeding, vigour is lost,' he wrote in 1929, 'and one cross will restore it ... It is a question whether an out-cross – that is, crossing of different strains ... will restore the vigour that may have been lost without having to cross two distinct breeds.'[31] The relationship of inbreeding to loss of vigour had concerned great poultry breeders like Stoddard and Felch and, even earlier, Sir John Sebright. There was nothing innovative about this information. Furthermore, comments on the value of hybrid vigour, on how to achieve it, and whether or not one should breed specifically for it and be prepared to drop purebred breeding had dominated the poultry press for at least thirty years. The apparently non-innovative aspects of selective methods put forward and the impracticality of much of the material presented by the scientists to breeders made many practical poultrymen confused about the implications of scientific research results. Geneticists also rarely acknowledged, at least publicly, the possible debt they owed to agriculturalists. (A good example of this phenomenon can be seen in M.A. Jull's discussion on inbreeding, provided in a 1932 edition of the *American Poultry Journal*, where he used Felch's chart and theory but did not credit the master breeder.)[32]

The advent of Mendelism in 1900 and the birth of genetics did not reduce the volume of discussion revolving around the Felch chart, inbreeding and outcrossing along the lines of Stoddard's utility breeding, and the double mating system.[33] Vague references to careful selection with little explanation of what that meant continued, but such discussions now often annoyed editors of the poultry press because of the lack of exactitude in the information. The idea that 'science' explained breeding methodology and selection, which Mendelism encouraged, inclined poultry breeders to put practical/craft breeding on a clearer (perhaps also a more scientific) basis. Long-standing emphasis on American Poultry Association standards made this difficult to do, because many breeders had become less sure what traditional breeding methods actually entailed. Reliance on the ethos of purebred breeding, which by this time suggested – but did not actually provide – a way to breed, did not help the situation.

At the meeting of the American Poultry Association in 1914, C.D. Cleveland (secretary of the New York Poultry and Pigeon Club, breeder, judge, and writer of many articles on chicken husbandry) commented on the poverty of material concerning selective breeding methods of craft breeders. 'I have never been able to understand why it was that the breeder was so loath to give away anything in regard to the essentials of the way he breeds his varieties ... Breeders ought not to hold back their so-called breeding secrets.' The editor of the *American Poultry Journal* thoroughly agreed, stating that 'Mr. Cleveland's remarks summed up the situation very nicely. We have been trying for years to get articles [on breeding] we want, and believe should be published, on the how and why of mating and breeding, but we can't get the information ... Perhaps some breeders may not be able to tell how they get results.'[34]

Craft Breeder Attempts, via Competitions and Standards, to Improve Chickens for Egg Production

Developing technology, particularly in the form of cold storage, triggered a greater focus by craft or farm breeders on the problem of breeding for productivity. The American Poultry Association considered standards based on productivity as early as 1903, when the committee in charge of revisions suggested that good utility should be recognized.[35] At the 1907 meeting of the association, it was resolved that 'the American Standard of Perfection [gave] undue prominence to the beauty value of standard-bred fowls, to the detriment of the utility value of domestic poultry.'[36] The association should organize standards for utility, just as it had for beauty. I.K. Felch agreed, believing that the American Poultry Association had been founded on the need for utility poultry and a professional breeding arm to serve the poultry industry, as he explained in 1910:

> The American Poultry Association is not doing what it should for the commercial end of the poultry industry ... Fancy poultry could not long exist without commercial poultry, and the commercial poultry is also dependent upon the fanciers to a great extent. Then why should we devote all our thoughts and energies to one side of the question, which we are in reality doing at present? The commercial poultry far exceeds the fancy in dollars and cents, yet we would not advocate doing less for fancy poultry, but undertaking more along commercial lines. We do not believe our work should be confined to shape and feathers alone, but to educational questions as to how to raise more and better poultry, and how to get the greatest financial returns for our investments should also receive due consideration at the hands of the American Poultry Association.[37]

Craft breeders had utilized the trapnest for utility breeding as much as had scientists. By the late nineteenth century it had become fashionable for craft breeders to trapnest certain birds in a flock, and in the twentieth century the device came to play an increasingly critical role in evolving craft-breeder efforts at standardizing utility breeding for eggs. The way the trapnest was used, however, caused clashes to erupt among breeders. The rise of egg-laying contests based on trapnesting lay at the heart of the matter. The first took place in Britain in 1897, when the Utility Poultry Club of England ran a competition, with only seven entries of four pullets each.[38] By 1912 egg-laying competitions had become common in the United States and Canada. All breeds competed, and contests were often dubbed the Battle of the Breeds.[39] Craft breeders used the contests to promote their stock. One ad in the *Farmer's Advocate* stated: 'My Barred Rock won the Canadian Laying Contest, Ottawa, laying 272 eggs. Cockerels and hens for sale.'[40] Breeders referred to Raymond Pearl's work when encouraging the sale of cockerels emanating from strains bred to lay, that is, from families where trapnesting contests indicated the good laying ability of a cockerel's female relatives.[41] 'It is a many times proven fact that the cockerel transmits the laying qualities to his pullets which he receives from his dam,' one ad reminded readers in 1920.[42] Trapnesting contests increased interest in maintaining records for poultry, and therefore an emphasis on pedigree keeping and genealogy.[43] They also introduced an element of sport to standardization, a factor which did not necessarily encourage better breeding practices.

Men knowing nothing about poultry bought from craft breeders in order to compete in the contests. Some did well, but the breeder, who deserved the credit, usually got none. If the birds did badly, the breeder might be blamed. 'Established breeders ..., of course, did not relish this state of affairs,' an experienced poultryman explained. Without these breeders, good egg records would never have existed, he argued. Many became embittered.[44] The conflict within the association between utility breeders over the issue of contests would be ongoing. Those who opposed trapnesting contests were often accused of being only interested in beauty. 'At the present time there appears to be a tendency, on the part of a few, to revive the old time warring between the ultra-fancy and ultra-utility poultrymen,' the *American Poultry Journal* noted in 1915. This is a mistake to fan the flames of an old feud. 'Instead of trying to warm over the old embers of contention between fancy and utility, it would pay much better to try to bring the beautiful and the practical into closer understanding and friendship for one another.' Breeders should get together and work together.[45] Conflicts over the value of trapnesting competitions within the American Poultry Association perhaps help explain why the association refused to organize and run such tests under its auspices.

Many craft breeders who opposed trapnesting contests were not, in reality, concerned with beauty. Nor did they reject the value of the trap-

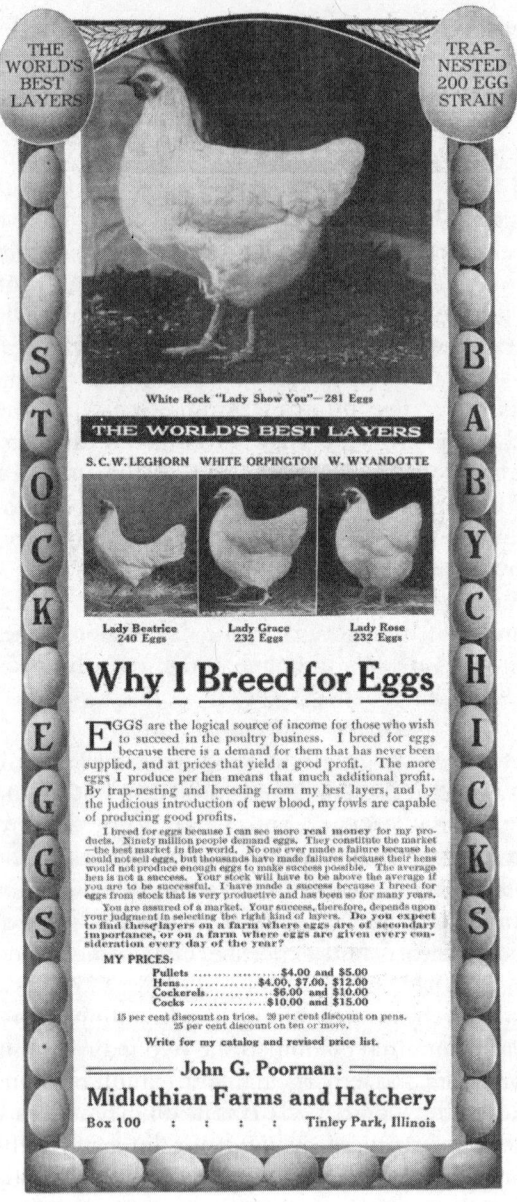

15 An ad based on egg-laying contests. Note the use of the dual-purpose breeds for the table-egg industry, and the emphasis on utility rather than exhibition qualities. (*American Poultry Journal*, December 1913, 1579)

nest for breeding programs. They simply thought competitions served no useful purpose. To begin with, they believed that records from contests were not accurately kept, and also that people running the competitions often had a personal interest in the outcome. The conditions for entries often varied too much as well.[46] All these factors undermined the credibility of results from contests, it seemed to them. They also saw the competitions as just that – competitions – and were not convinced that competitions led to good breeding practices.[47] In fact, they argued the reverse. Much of their thinking was in line with the geneticist point of view concerning the use of the trapnest. For many craft breeders and geneticists, trapnesting competitions encouraged the idea that breeding should proceed via individual worth. The crux of the matter was that, while trapnesting competitions might identify what birds were heavy layers, that fact did not explain how to breed birds for such a character. The purpose behind the test should not relate to competition if it was designed to improve breeding, both craft breeders and geneticists argued. Within that framework, volume was also important: a large number of fowl, randomly selected, were needed. These convictions, shared by certain craft breeders and geneticists, indicate how similar genetic and craft breeding principles could be. Similarities of this nature fed into the on-going problem of distinguishing scientific breeding from craft breeding.

Government Regulation: The Record of Performance and Breeding for Egg Production in Canada

Another factor soon played a role in this confused environment of craft breeders, namely, the authority of the state. When governments decided to encourage better breeding, the culture of the day dictated recognition of breeder efforts and expertise. Governments logically turned to structures supported by breeders and poultry experts at agricultural colleges.[48] State orchestration of chicken breeding, therefore, came to be based on a system that combined the two main organizational tools of the craft breeders and experts, namely, exhibition standards and the testing of egg-laying capacity via trapnesting practices set up in contests. The new government system also intensified the beauty/use dichotomy by uniting beauty standards with production standards. That situation ultimately brought the long-standing beauty/use conflict to a head, and in the end forced a parting of the ways between the two concepts. But not before the craft/science divide became entangled with the beauty/utility divide in the minds of some craft breeders, thereby playing into the difficulty (then and now) of separating the way scientific breeding differed from craft breeding.

State regulation of poultry breeding in North America began before the First World War and was initiated first in Canada, a country that had used governmental authority to regulate livestock breeding organizations in a way less likely to be seen in other Anglo countries.[49] The Canadian government, for example, regulated livestock associations after 1900 and soon certified Canadian pedigrees of purebred livestock – horses, hogs, and cattle – under the Livestock Pedigree Act of 1912. No such act existed in either the United States or Britain, and livestock associations functioned entirely independently in those countries. William Graham of the Ontario Agricultural College initiated the move in Canada towards state involvement with respect to chickens, specifically the organization of poultry breeding for egg layers. He had watched with interest developments in the North American dairy-cattle industry and the new attention to milk-yielding performance that breeders gave over the last part of the nineteenth century, and that government did at the beginning of the twentieth century.

Since the 1870s Jersey cattle breeders, and the 1880s Holstein Friesians breeders, in the United States and Canada had collected milk-production performance data on dairy cows, and used the data in breeding selection. In 1894 Holstein breeders in North America began official supervision of recording on the basis of milk performance – bulls were admitted too, on the basis of the milk production of their daughters. The Holstein Friesian Association of Canada established a system of Advanced Registry in 1901 with its Record of Merit (ROM). It was a short term test (seven to thirty days) of milk yields. In 1905 the Canadian Department of Agriculture initiated a state-endorsed test, which ran for ten or twelve months of the year and was known as the Record of Performance (ROP). It was available for all cattle breeds. (Holstein breeders began entering their cows in the ROP in 1907. Shorthorn breeders followed in 1913.) Cows were tested on the farm and inspectors made regular checks.[50]

As early as 1912 Graham suggested a way to breed for a higher standardized egg-laying capacity in flocks, based on the collection of egg-laying data, done by the Canadian government and run in conjunction with regulations relating to the Standard of Perfection. Graham pressed more actively for a poultry performance plan similar to that for dairy cattle at the first Canadian National Poultry Conference held in Ottawa in 1919. He stated that if the federal government would not undertake to run a Record of Performance to label superior egg-laying lines in standardbred poultry breeds, then the province of Ontario would do so within twelve months.[51] A Dominion government committee, accordingly, was set up to look into the matter.[52] By the summer of 1919 regulations for a national Record of Performance had been established. In order to qualify for the Record of Performance a bird had to lay at least 150 eggs in 52 consecutive weeks on the farm and the trap-nesting records would be checked by inspectors. At least 10 birds were needed for each entry in the test. To qualify for the Advanced Record of Performance a minimum of 225 eggs were required in 52 consecutive weeks, and the test had to take place at the Dominion Experimental Farms, or at provincial department or college grounds. Only birds of standard varieties and free from disqualifications outlined by the American Poultry Association in the Standard of Perfection were eligible for acceptance under either type of ROP certification.[53] The ROP was firmly attached to traditional views concerning the value of purebred, or standardbred, breeding over that of mongrelization because of its linkage to the Standard of Perfection.

The Canadian ROP provided breeders across the country with the stamp of approval of the national government, and quite quickly the provinces set up associations to organize testing within their territory. While Ontario (under Graham's pressure) might have been the chief instigator of a federal ROP system, by the mid-1920s the province that produced the most ROP birds in Canada and had the highest percentage of particularly superior egg-laying hens was British Columbia. Interest in breeding for egg laying and in the ROP itself was encouraged in that province by the work of a strong poultry-breeding research program (done under the direction of E.A. Lloyd and V.F. Asmundson) at the poultry farm of the University of British Columbia. Lloyd organized the British Columbia ROP Association, and encouraged private breeders in the province to follow breeding methods established at the college to produce superior ROP birds. British Columbia breeding stock received international acclaim. The success of the ROP in British Columbia and the general poultry research being undertaken at the University of British Columbia commanded the considerable attention of American breeders. In 1926 a delegation went to Vancouver to look into the dynamics of the Canadian ROP in that province. There was no question in the minds of the visiting Americans that the ROP's connection to the Standard of Perfection was intimate. As A.D. Duston reported to the *American Poultry Journal*, the government inspectors of

ROP entries were instructed to examine every bird for disqualifications set by the Standard, and to do so with rigour.[54]

The Canadian ROP structure, because of its essential connection between productivity and recognized standards of beauty, encouraged cultural shifts in attitudes to chicken breeding. Adherence to purebred breeding principles, present from the time of Burnham in both the United States and Canada and therefore even before the advent of the Standard of Perfection, fit in well with the ROP, which encouraged the record keeping of stock so qualified. In 1923 the Canadian government provided for the public recording of ROP poultry.[55] Birds eligible for recording in the Canadian Livestock Records Corporation had to have laid 200 eggs or more over 52 consecutive weeks in the Canadian Egg Laying Contests, averaging a weight of 24 ounces per dozen, and be free of disqualifications set by the Standard of Perfection.[56] Recording, of course, encouraged the growth of a system of pedigree keeping, and also a reliance on ancestry breeding. (The association representing ROP breeders operated independently from the association that spoke for the interests of recorded purebred poultry, the Canadian Poultry Record Association. The overlap of concerns of the two had always been considerable, and in 1938 they amalgamated.)[57] Breeders using ROP stock or birds recorded in the Canadian Livestock Records Corporation tended to breed from such registered birds, thereby relying on ancestry breeding. The recording of hens in Canada was ultimately based on a complicated interplay of standards (via the ROP), egg-laying contests, and ancestry.

Breeders and government officials in both Britain and the United States noted Canadian ROP developments. The editor of a British journal, C.A. House, wrote the following: 'The accumulative effect of the ROP and registration is to give Canadian breeders a system of pedigree breeding, trap nest inspection, and progeny testing together with dependable records both of quality of stock and egg production that no other country possesses.'[58] Americans were even more impressed.[59] 'The registration of chickens in an official record book, so as to afford an authentic basis for tracing pedigrees, is something new in the poultry world, although it has long been established' for other livestock, stated the *American Poultry Journal*.

> Now comes the Canadian government with the announcement that it has granted to poultry breeders the right to register purebred chickens in national registration records. This is a step toward 'blue blood' and a

'poultry aristocracy.' The National Poultry Record Association has been organized in Canada and has opened a registration book to all Canadian breeders of purebred fowls. Mongrels cannot be registered.[60]

The credibility of ROP breeding was greatly enhanced by government sanction.

Government Regulation: The Record of Performance and Breeding for Egg Production in the United States

A move towards an American Record of Performance developed fairly quickly after the Canadian structure was established, and these were modelled on the Canadian one. The reasoning behind ROP regulations in the United States, however, was somewhat different. The issue of falsified egg-laying records seemed to plague American craft breeders more than Canadians, and that problem, more than attempts to direct breeding, drove support in the United States for use of an ROP.[61] It was believed that some form of state regulation of trapnesting tests and contests was needed to ensure the collection of honest records. 'As we look at it, the ROP [in Canada] has blazed the trail in the matter of a complete system of registration of poultry,' the *American Poultry Journal* explained, adding that recording and inspection were truthfully done.[62] 'The honest breeder, working at home on his own plant, should be able to command public confidence,' another American stated. 'But, unless the US Government, patterning after the Canadian system, shall arrange to send an inspector to make unannounced examination of flocks that are on test, we know of no way to combat "home grown" egg records, and the evil must continue unabated, and with increasing velocity, until it breaks up from its own fierce momentum.' In Canada, officially kept records were recognized as being legitimate, the journal told its readers.[63] Confidence in the better quality of the birds also created a demand for the stock in the ROP, poultry experts in the United States noted.[64]

The American Association of Instructors and Investigators in Poultry Husbandry authorized in 1919 the initiation, under its auspices and with its moral support, of the American Record of Performance Council, which was intended to introduce Record of Performance principles, developed along Canadian lines, to the United States in order to check the dangers of false egg-laying data. This national body was put in place in 1921, but not until 1924 did the American Record of Performance Council begin to deal with 'the persistent problem of certifying egg-production performance records in some sort of official manner.' Competitions, however, continued to play a role in how good egg-laying birds were identified for ROP status in the United States, as was true in Canada. The council was made up of all those running egg-laying contests in both the United States and Canada who wished to join. Pullets laying two hundred eggs a year qualified for recording.[65] Unlike the situation in Canada, ROP regulation in the United States tended to be more regionally or state-oriented than national in nature. The national body did not seem to draw as much support as local state organizations that established egg-laying contests. State ROPs had already come into existence. Washington State and Massachusetts each had one by 1923. Iowa and New Jersey joined the ranks in 1925.[66]

The council's inclusion of Canadians is interesting. American and Canadian chicken-breeding interests were still closely linked. The North American breeder association, the American Poultry Association, represented breeders in both Canada and the United States, a fact well recognized by American poultrymen. The commonality of breeder affairs and joint organizational structures for poultry was unusual in the North American livestock world. Various cattle and horse-breed associations were normally nationally oriented. And there could be considerable friction between Canadian and American organizations, as affairs in Shorthorn cattle and Clydesdale and Percheron horse associations make clear.[67]

By 1930 a more viable American national ROP system based on set standards outside contests had come to exist. Some sixteen states joined together to form the new United States Record of Performance Association, which recorded egg production, egg weight, and body weight made on the breeder's premises under official trapnesting supervision for a year. The regulations matched those of the Canadian ROP standards. Each pullet candidate of the USROP was banded. Females that qualified for the USROP laid two hundred eggs or more during the trapnest year, such eggs to average a minimum egg weight of 24 oz per dozen. Each bird had to be a good representative of its breed under the Standard of Perfection. The highest breeding stage in the scheme was the United States Register of Merit (USROM).[68]

Poultry ROP work in the United States never commanded the respect from utility craft breeders that it did in Canada. Some Americans believed that government interference with breeders and added breeder expense were the likely outcomes of a national ROP system. Note the contemporary comments by a poultryman, J. Miller, who wrote in to the *American Poultry Journal* on the subject:

> An effort is again being made to form a registry association for poultry [in the United States], and, as before, it seems that those supporting the proposition are not breeders themselves, but are persons who are very anxious to dictate the work of the breeders ... The plea is being made that registration will be the means of improving the quality of poultry in general, but this argument is without foundation, because there is no possible way by which registration can or will add anything to the quality of the bird. It will add only to the cost of producing and selling the bird. Improvement is made by careful, intelligent breeding, not by writing leg band numbers in a register.[69]

The effect that the Record of Performance had on breeding principles also alienated American craft breeders who had opposed the contests for breeding reasons. The ROP was not based on the type of thinking that underlay support for competitions. It was, rather, based on the idea that all birds meeting a preset standard should be recorded. However, emphasizing a standard in this fashion encouraged many of the undesirable breeding habits that the contests did, namely, a reliance on individual hens and the promotion of ancestry breeding.[70] Inadequate volume was also a problem in the early ROP structure. (The advent of the random-sample test for layers in the 1950s brought better and more telling results than the older standard test that governed the ROP, where the owner selected mature birds that he wanted to enter from his flock. In the random test, a disinterested person picked entries at random, and could even start with unhatched eggs. In the standard test an entry consisted of roughly twelve individuals per owner. In the random test an entry numbered from fifty to several hundred.[71] While the standard test could have worked with progeny testing, the orientation towards individual hens did not lend itself to that approach. Lack of volume per entry was another problem. The random test could be used to progeny test more easily than the standard test. The use of the random test was suggested as early as 1927 by a Dutch geneticist, A.L. Hagedoorn.[72])

In light of the conflicts over an ROP for poultry, it is perhaps ironic that the record-of-performance system not only continued to prevail in the dairy-cattle and poultry industries, but also expanded to affect other American farm animals. Efforts were being made as early as the 1920s to develop a record-of-performance scale for beef cattle in the United States. Some beef-cattle schemes were based on show-ring success and progeny, while others took growth rates, ability to utilize feed efficiently, or carcass grade as the standards for a record of performance. It was believed in the 1930s that one of the problems with all such systems was that they relied on male stock, namely, bulls and steers, and therefore took no account of the genetic contribution of females or breeding cows (which, because of variability in any cow herd, was difficult to do). Record-of-performance systems for swine in countries outside North America predated those for other livestock, including poultry. ROP regulations for swine were in place in Denmark as early as 1907, Germany beginning in 1926, and areas of Britain from 1927. In 1926 American experiment stations in Iowa, Minnesota, Wisconsin, and West Virginia began to work with the Danish ROP system, but success with these projects was limited. In 1934 a swine specialist representing the Iowa experiment station went to Denmark to find superior stock that had been tested under the Danish system, and to learn about feeding methods. An American ROP for swine was still in the future, but concerted efforts to bring it about were farther ahead than in the beef-cattle world, and cross-breeding for terminal results along the hybrid-corn method was also in its initial stages with respect to swine.[73]

Generally speaking, it was the most significant American poultry breeders for the future of chicken breeding who opposed using the Record of Performance. Often their breeding methodology, which demonstrated a reliance on statistics and an emphasis on populations rather than individuals, matched that of geneticists. The affinity of their thinking with geneticists in this matter tended to ally such breeders with the new science. Even so, it is not always evident how closely such breeders worked with geneticists. Some breeders seemed to be influenced most by what experience, not science, had taught them. They simply drew the same conclusions as geneticists did in such matters, but they did so independently. Parallel thinking of this nature makes it hard to see which influenced the other.

The Influence of Geneticists on Significant Early Commercial Breeders

It is not clear how much (or if) the work of M. Johnson in Texas was influenced by geneticists when he abandoned the breeding of exhibition poultry in favour of focusing on selection within large breeding populations for utility features. As a boy he had raised exhibition chickens, and after marriage in 1904 he continued to keep birds, but now for commercial reasons: he needed the extra income they provided in the form of eggs, and stock sold for breeding purposes. By 1908 he had decided to rely on sibling and progeny testing from his own trapnesting records for the mass production of utility poultry. He built a composite strain from three distinctly different ones within the Leghorn breed, and developed a breeding pool made up of thousands of pullets, as well as generating eggs for sale on a large scale. His views on poultry keeping went hand in hand with the increasingly prevalent belief that agriculture should be industrialized through mass production. The poultry press extolled this virtue in 1929, stating that Johnson followed the factory methods of Henry Ford.[74] Johnson promoted his factory-based operations by stating that his ranch had managed to turn out literally thousands of good utility breeding and producing birds every year. His 1927 ad announced that he produced the best strain of egg layers in the world.[75]

It is completely evident that other early breeders communicated with scientists working at experiment stations. D. Tancred, for example, who depended heavily on statistics generated by large numbers of birds, acted on the advice of G.M. Gowell (who preceded Pearl) at the Maine Experiment Station by using a type of trapnest devised there. Tancred relied on sibling or progeny testing, rather than ancestry breeding. Birds of the Tancred strain of Leghorns remained popular with farmers until well into the 1920s.[76] Johnson favoured using Tancred lines in his breeding program. Both Johnson and Tancred entered egg-laying contests for publicity reasons. But, significantly, they did not use the results of the contests for breeding purposes.

Another breeder who had a close working relationship with a poultry scientist at an experiment station was J.A. Hanson. Educated in agriculture at the University of Missouri, Hanson moved in 1911 to Oregon to work under James Dryden, who ran the poultry department at the Oregon Experiment Station from 1907 until the 1920s. In 1913 Hanson set up his own breeding operations, utilizing lines that Dryden had developed at the station. Hanson carried several independently bred strains which he combined to produce a commercial bird.[77] His system resembled the future heterosis breeding of geneticists, but it had an affinity as well to inbreeding and crossing systems designed by Stoddard.

A few utility breeders chose to rely completely on scientists and geneticists. One early utility breeder interested in the work of geneticists was E. Parmelee Prentice (1863–1955). A Chicago lawyer, he bought a country estate, Mount Hope, in Massachusetts in 1910. Interested in increased food production and excited by the potential of Mendelism for better farming, he decided to go into agriculture.[78] 'When Mount Hope, in 1910, began its work,' he explained, 'its attention was ... given entirely to the development of methods for increasing the production of food. The greatest possibilities of such increase, it seemed to us, lay in seed and plant improvement and in raising the productive qualities of our races of domestic animals.' Particular attention was given to dairy cattle and egg-laying poultry.[79] Prentice hired a cattleman to buy top producing Guernsey cattle. The cattle might have been beautiful, but they provided low amounts of milk. Shocked by this lack of productivity from the best purebred dairy cows, Prentice founded a new registry association for cows that gave at least 4000 pounds of butter fat a year. Cows eligible by milk yields did not need a pedigree in a purebred herd book, and their colour was of no consequence. He went on to write a book on the breeding of dairy stock, *American Dairy Cattle: Their Past and Future*.[80] After 1922 H.D. Goodale, poultry scientist at the Massachusetts Experiment Station, ran the Mount Hope cattle- and chicken-breeding operations on a full-time basis. A specialized poultry staff worked under Goodale.

Prentice had enduring success with his White Leghorns, which went on to play a major role in the modern breeding business of table egg hens. Goodale's breeding efforts resulted in superior egg-laying birds, and by 1950 Prentice believed that these hens should no longer be called White Leghorns, a breed that was established under the exhibition system. 'For thirty years they have been segregated from all other poultry,' he explained,

> and inter-bred for the increase of the productive inheritance. During this time no attention whatever has been given to their appearance. We do not know how judges of exhibition poultry, familiar with the *Standard of Perfection*, would regard the web, fluff and quills of their feathers, or their other properties, qualifications or disqualifications, which, according to the *Standard*, White Leghorns should possess or should not possess.

Prentice's poultry-breeding operations spanned the period which saw the decline of the exhibition-bred bird as commercial stock and the rise of the geneticist as practical breeder.

John Kimber, with a degree in agriculture and music from Stanford

University, began breeding Leghorns in 1925. He bought his foundation stock from Dryden and Hanson, and consulted with geneticists on how to establish a suitable research program in support of his breeding operations. Along with James Dryden, he adamantly and openly opposed the movement in 1930 to form the United States Record of Performance Association. By that time Kimber had moved decisively over to quantitative scientific breeding in conjunction with the progeny test. In 1943 he hired a geneticist with a PhD on a full-time basis.[81]

The Genetic/Craft Conflict: Differences between Canada and the United States

As early as the 1920s American breeders who continued to work under the American Poultry Association's Standard of Perfection found themselves in an increasingly difficult situation. Virtually no one interested in agricultural production was concerned at any level with beauty. In 1923 the geneticist L. Cole told a group of breeders that there were two possible paths to follow in order to solve this divisive problem: either separate utility breeds from fancy breeds and 'acknowledge them as such,' or make the standards for fancy points much broader.[82] The *American Poultry Journal* believed Cole's fundamental premise that utility breeds and beauty breeds should be clearly separated from each other was the right one to follow. The demand for utility was so strong that promotion of beauty in utility breeds should stop. 'The demand today is for useful stock,' the journal pointed out. 'Utility breeding is gaining ground. It is a period of transition ... This is not the day of art. It is the day of commercialism.'[83]

The American Poultry Association refused to change its position about the importance of beauty to any breeding program, and, as a result, encouraged many American craft breeders who wanted to focus on utility to abandon the Standard of Perfection as a breeding guide. In spite of pleas for recognition that times had changed, some older fanciers were not prepared to admit that the hegemony of standardbred breeding was nearing its end as far as the commercial poultry industry was concerned. Increasingly it began to appear to many breeders that the ongoing marginalization of standardbred breeding resulted, not from a confrontation between beauty and utility interests or divisiveness among craft utility breeders over trapnesting use, but rather from what often seemed to be clashing approaches to breeding between craft and science.

E.B. Thompson, an American breeder who attended the American

Poultry Association's meeting held in Toronto in 1924, claimed that the standardbred system for poultry breeding had been assassinated by the press and the agricultural colleges, both of which ridiculed craft breeding generally and at the same time supported genetic research in breeding. Canadian colleagues were astonished at the depth of the cleavage between craft and science in relation to breeding methodology, and at the contingent threat of science to craft breeding that apparently existed in the United States. They did not see genetics as being in competition with traditional breeding. At the meeting they pointed out that the American Poultry Association was the only breeding organization recognized by the Canadian government. William Graham described his twenty years' work with Barred Rocks, selecting for greater productivity but within the framework of the Standard of Perfection. Canadians believed that utility breeding could be done within the existing structure.[84] The problem for Canadians was overcoming the beauty/utility divide by focusing on breeding for utility and incorporating new ideas arising from science to that end. They saw no need to abandon traditional organizations, or good selective-breeding strategies long understood by craft breeders. The more general Canadian craft-breeder support of an ROP encouraged this contentment with the status quo.

There still were American craft breeders who also believed the status-quo approach could serve the commercial industry, and who did not see genetics as the enemy of standardbred breeding. T. Hewes, a man who wrote articles for various poultry journals, reacted to Thompson's comments (which suggested the death of traditional breeding and the take-over of genetics) by stating that, as far as he was concerned, the standardbred bird was anything but dead and anything but useless for farm purposes. 'Edward Thompson's speech at Toronto reads like a child's plea,' he said, adding 'One would think the end had come and the whole Standard poultry business was only waiting for the undertaker.' The problem was not a division between science and craft breeding approaches, Hewes argued; it was the old division between beauty and utility orientation that was so detrimental to standardbred interests. Craft breeders who moaned about the future were too focused on beauty, and not on breeding to provide useful stock with the aid of trapnesting records, Hewes declared, and expanded as follows:

> The trapnest never hurt any breeder except the one that failed to install it. The wise fancier put them to use work years ago and they proved their worth many times over. If you don't know what your hens are doing, if

you are not keeping any record to show whether or not there is a profit commercially in the business, then you are a sloppy housekeeper and in no way entitled to sympathy ... Get this fact thoroughly fixed in your mind, the poultry business of this country was built on Standard bred birds.[85]

While for this breeder the beauty-versus-utility conflict, not craft-versus-science tensions, explained why standardbred birds received increasingly less respect from the commercial poultry industry, it is apparent from Hewes's views that utilization of the trapnest played a role in breeder disagreements. How to use it in order to proceed with utility breeding, however, was by no means clear from his words.

The 1924 meeting of the American Poultry Association revealed that, regardless of internal, divisive craft-breeder problems, the competition between scientist and craft orientation to breeding was stronger in the United States than in Canada, and that geneticists played a more significant role in research breeding in the former country as well. The early hiring of geneticists at American experiment stations played a role in both phenomena. The linkage of trained geneticists with agricultural institutions, even if the crossover of academic science to practical use remained limited, was not common outside the United States at that time. The Dutch geneticist Hagedoorn argued that by 1930 there were not as many as twelve geneticists working in any form of agricultural animal breeding in the world outside the United States.[86] The Canadian situation mirrored that non-American international trend. Trained scientists working in poultry breeding at agricultural colleges were the exception rather than the rule. One important breeding centre that utilized scientific training arose at the University of British Columbia, where a poultry department was established in 1917/18. E.A. Lloyd, a graduate in agriculture from the University of Saskatchewan, and V.F. Asmundson, also a graduate of Saskatchewan and Cornell as well, established influential breeding lines in the 1920s, but it is unclear how much knowledge arising from genetic research shaped this early work. At any rate, efforts at breeding in British Columbia remained in line with traditional thinking, which supported the establishment of strains that bred truly and the use of the ROP.

In the 1930s, the Depression terminated much of the breeding research being done at the British Columbia agricultural college, when the university greatly reduced its poultry work. Asmundson relocated at the University of California. The reduction of poultry-breeding work at the University of British Columbia led an important future geneticist,

I.M. Lerner, to move to the United States and relocate at the University of California as well, where he received his PhD in 1936. Lerner experimented with poultry most of his professional life, and while he contributed greatly to an understanding of how selection practices worked in breeding for egg laying, like so many of his contemporaries at agricultural experiment stations, Lerner's main interests continued to be the mechanics of inheritance and evolution. His poultry experiments serve as an excellent example of how poultry-breeding research could until the late 1930s lend itself to the on-going basic dichotomy: namely, between poultry genetics as a way of understanding genetics and evolution and poultry genetics as an avenue to greater agricultural productivity.[87] Scientists often seemed to see one as synonymous with the other, a situation which only added to the confusion of breeders who read reports of research that studied both in the farm press.

In Canada early breeding experiments tended to be done by poultry experts who had no specific training in either biology or genetics, W.R. Graham being the prime example because of his importance to the poultry world of Canada and because of the position of the Ontario Agricultural College in Canadian agricultural education and research. Graham ran the poultry department at the college from 1899 until 1940 and was informally educated at best. A superb breeder, Graham believed in a utility emphasis. His Barred Plymouth Rocks bred at the college found immense favour in Canada and a number of northern American states. Housing, nutrition, and artificial incubation took as much of his attention as breeding. He appreciated the work of scientists working in all poultry-related fields, remaining a friend of Pearl, for example, throughout his life. Graham was important, not just to the Canadian poultry world, but to the international one as well. He, along with J.E. Rice, who ran the poultry department at Cornell University, was the main driver in the earliest international organizations which met to discuss poultry affairs, and he commanded worldwide respect. Graham and Rice founded the Poultry Science Association in 1908, which drew together poultry experts across the United States and Canada, and were instrumental in the organization of the first world congress on poultry affairs held at The Hague in 1921. Graham also loved beauty in the birds, and qualified as a show judge under the American Poultry Association. He was asked a number of times to judge important exhibitions in the United States.[88] Graham's affiliation with both utility and fancy factions, his ability to breed well in spite of his lack of genetic training, and his prominence in chicken affairs helped defuse tension in Canada between the craft and science outlooks to breeding.

Graham trained a number of poultry experts who became geneticists or scientists in other fields, many of whom ended up in the 1920s and 1930s at American colleges, experiment stations, or the USDA. A particularly important example was F.B. Hutt (with a master's degree from Wisconsin, a PhD from the University of Manitoba, and a subsequent staff position at Cornell University). Canadians interested in poultry affairs, in fact, frequently moved to the more robust scientific environment of the United States. M.A. Jull (with a degree from the University of Toronto and PhD from Wisconsin), held poultry positions at the West Virginia Experiment Station, with the USDA, and subsequently at the University of Maryland. While poultry experimental breeders in Canada on the staff of agricultural colleges were not geneticists in this period, they did not seem to feel that genetics undermined their position or the traditional way to breed. The fundamental threat to good chicken breeding for them came from an overemphasis on beauty, not science.

In many ways the ROP was the most significant approach for utility production offered by craft breeders. The endorsement of government enhanced their stance, even in the United States. But the ROP aggravated the inherent tensions in the traditional chicken-breeding world by dividing utility craft breeders into two camps, particularly in the United States. The ROP promoted breeding by individual and ancestry, and also tied utility breeding to breeding for beauty. In order to qualify for the ROP a bird had to meet the Standard of Perfection. Perhaps equally important, the ROP system masked traditional knowledge on how to breed, because by the late 1920s the ROP had come to be seen as a craft-breeding methodology itself. Articles devoted to breeding methods that were based on tradition appeared less frequently in the poultry press, a pattern quite evident by 1930. Virtually no articles on the subject were printed after 1930. Increasingly, fewer people could describe craft breeding, and therefore compare it with genetic breeding. By the 1930s it was the ROP, not craft-breeding methodology, that competed with genetics for attention from utility breeders.

Levels of Egg Production

Despite the work of craft breeders and geneticists, and also the husbandry efforts of producer/growers, improvements in egg production did not appear to be stunning between roughly 1900 and at least 1930.

Figures given for numbers of eggs per hen in the period were somewhat inconsistent, making it difficult to estimate actual egg-laying levels, but certain trends seem evident. One study set the annual production of a hen at 86 eggs per year in the United States in 1913,[89] but the *American Poultry Journal* reported seven years later that the US census of 1920 indicated a level of only 55 (4.6 dozen) eggs per hen. The US census of 1930 found egg production had risen to a mere 69 (5.75 dozen) eggs per hen by 1930.[90] (The Canadian government's figures, which stated that hens on average laid 86 eggs a year in Canada in 1924, seem questionable, especially in light of American data, varying as it might be.)[91] The USDA *Yearbook* for 1943–7 suggested that by the 1940s the average hen egg-laying rate was 118 eggs per year.[92] It would appear that any real advancement in egg production did not start before the 1940s. Even then, though, it is difficult to argue that improvement resulted from superior breeding. Advancements in nutrition, health, and housing were significant in the period, accounting easily for the better performance of hens. The poultry press, in fact, devoted more attention in the 1930s to developments in these areas than to any form of advancement in breeding techniques.

For a number of reasons, the period between 1900 and 1930 saw an increasingly heightened conflict, particularly in the United States, between craft breeders of poultry and geneticists over how to select for better production, generally with a focus on the egg industry. A particularly important issue separating American geneticists from craft breeders needs to be pointed out. As early as the 1920s there was an increasing tendency for geneticists to link systems of inbreeding with hybridizing in chicken breeding, and in doing so the scientists advocated the breeding of commercial birds that could not reproduce themselves truly. This approach was diametrically opposed to that of craft breeders who, while they might use inbreeding strategies, preferred to seek improvement through the creation of strains that bred truly. On top of practical problems relating to volume and a fundamentally different theoretical orientation, geneticists had provided little evidence yet that, by breeding hybridized, commercial stock that would be unusable for breeding purposes they could produce superior birds with any consistency, in considerable volume, or at a reasonable cost to the producer/grower. The situation was made more contentious by other factors: divisive cleavages within the craft-breeding world, the clouding of methodology by organization and regulation, and the gendered division between breeder and producer. The organization of genetic poultry breeding would change profoundly over the late 1930s and 1940s, however, a situation which created insurmountable challenges to the old way of breeding. The resolution of the craft/genetic conflict in breeding began in the egg industry, and the way the hatchery industry orchestrated the breeding of egg-laying hens played an important role in how that unfolded. The story of the hatchery industry in relation to breeding for egg layers is the subject of the next chapter.

Chapter Five

The 'Scientizing' of Breeding in the North American Egg Industry

One of the most important technological developments that impacted the poultry-breeding industry was the invention of workable incubators which supported hatchery operations. When it became feasible to hatch chicks artificially in large numbers and when it was understood that a baby chick lived on its internal yoke for at least 72 hours after hatching, a new industry, the hatchery industry, was born. Its connection to the breeding industry was intimate over the first half of the twentieth century. As one prominent twentieth-century breeder put it: 'The hatchery industry developed from the breeder industry.'[1] Between the mid-1920s and about 1940 the hatchery industry in North America in effect directed the egg-breeding industry, whether hatcherymen themselves were breeders or not. The hatchery industry also encouraged the rise of new positions between breeders and producer/growers, and played a significant role in a trend to vertical industry integration. In this chapter, I focus on the North American hatchery industry and its relationship to egg breeding from roughly 1900 to 1950.

Artificial Incubation and the Rise of a Hatchery Industry

Artificial incubation had been practised for thousands of years by the time the first commercial incubators came on the market in North America near the end of the nineteenth century. It was well known that Egyptians artificially incubated chicks, and methods for performing this feat with home-made contraptions were detailed as early as 1732 in Britain.[2] In 1750 a commercial incubator was invented in France. In 1844 an Englishman patented another one. By 1880 the artificial incubation of chicks by commercial operators was common in both Britain and France. Breeders brought their eggs to these centres, and the incubator operators hatched the chicks, which would be picked up by the breeder within several days.

In 1873 the first incubator in the United States was patented by J. Graves of Boston.[3] Incubators were available in Canada by 1875.[4] Some sort of commercial hatchery operation was functioning by 1880 in New Jersey.[5] One also existed in California by this time. The incubating of chicks and the concurrent shipping of them would shortly follow. By 1892 J.D. Wilson of New Jersey was operating a commercial hatchery and selling baby chicks. That year he shipped baby chicks to a poultryman in Illinois.[6] The first shipment of baby chicks in Canada – from Ontario to Quebec – took place about 1900.[7] An Ontario poultry breeder, W.A. Fisher, was unaware of these North American shipments. In 1905 he read in an English newspaper that day-old chicks had been shipped by E. Brown from England to France for some years. He decided to experiment by putting twelve newly hatched chicks in a cardboard box and leaving them alone for three days. When Fisher opened the box he found all chicks alive and well. He began to advertise the sale and shipping of day-old chicks, and received orders from as far away as Winnipeg, Manitoba. His five-hundred-egg incubator worked to capacity to fill his orders.[8] By 1913 day-old chicks were making the long trip by train from Ontario to the foot of the Rocky Mountains.[9] The ability to move chicks over considerable distances meant the ability to transport superior genetics to destinations many miles away from where they had been generated. Hatcheries quickly improved the way genetic material could reach farmers across the United States and Canada.[10]

Small incubators were readily available by the early twentieth century, and these were designed for on-farm use by producer/growers, where sitting hens would be replaced by the machines. The technology made it easier for a farm to raise the non-sitting egg producer, the Leghorn. Women often liked the machines, because they reduced the poultry workload.[11] In 1903 one woman, in an article entitled 'Incubators versus Hens,' reported using an incubator for three years. An incubator was a necessity, she argued. 'I don't think I am an exception, but only one of a large number of farmers' wives who have more work to do than they can well manage.'[12] Capacity was small – no more than a setting for five hundred eggs. By 1905 ads in the Canadian farm press focused on selling these small incubators to farm wives.[13] Farm incubators, however, did not solve chick problems. Available brooders often could not handle the volume of incubator-hatched chicks.[14]

In spite of these difficulties, incubator manufacturers continued trying to generate sales from producer/growers, but as early as 1909 it seems clear that most Canadians who owned small incubators were breeders, not producer/growers. Breeders sold baby chicks, and sometimes also custom-hatched for farm families.[15] Increasingly after 1912 the *Farmer's Advocate* carried more ads for day-old chicks than for small incubators, indicating that producer/growers were more likely to buy baby chicks. A review of ads in the farm press up until the 1920s indicates that this pattern continued: it was the breeders who used small incubators, not producer/growers. (Breeders, too, often turned to custom-hatching at other hatcheries.) The breeding industry merged into the hatchery industry in this fashion.

By the 1920s virtually all Canadian producer/growers who wanted hatchery-incubated chicks bought the baby birds off the farm, or else took their eggs to a commercial hatchery for incubating. 'Why bother hatching when you can buy at 27 cents each?' one ad told readers in 1922.[16] 'There are many who do not care to bother looking after an incubator,' the *Farmer's Advocate* explained in 1923, 'but yet they want early chicks. This trade is being catered to by a number of poultrymen, who make a practice of selling day-old chicks. This is a very handy way of starting or replenishing the flocks.' Costs are small, the journal added, and the risk minimal.[17] 'If you can get your own eggs hatched not too far away from home, at a cost of two or three cents per egg, it would, under most conditions, be better for you to have them hatched than to buy a machine, and hatch for yourself.'[18] It should be pointed out, however, that hatcheries had not taken over the job of incubating eggs by the 1920s. Sitting hens continued for some time to be widely used on farms to hatch chicks. William Graham of the Ontario Agricultural College estimated that by 1932 about 40 per cent of chicks in Canada were hen-hatched, 40 per cent were hatched by large incubators, and 20 per cent by small incubators, although he was forced to admit that no one actually knew whether these numbers were correct. Large 'commercial poultry plants' also relied on incubators to hatch chicks in huge volume, but such operations were extremely unusual in 1912. More common by 1932, they still did not represent much more than 10 per cent of production.[19]

It tended to be breeders who bought small incubators in the United States as well by 1910. Sometimes women used the machines, selling day-old chicks of their own breeding.[20] In 1907 a woman advertised day-old chicks from her high-producing Leghorns and Barred Rocks hatched at her Sunflower Hatchery.[21] Breeders raising exhibition birds bought incubators too, although many found the issue of artificial incubation contentious. In 1910 the *American Poultry Journal* advised all breeders to overcome any disdain for the machines, and to get one. 'The breeder or poultryman who is up and doing knows what a boon the incubator is to him. If he [wants] hen hatched chicks he may get his eggs started well with the machines and then later put them under hens as they go to setting. Generally speaking, [even small breeders] need a good machine to get out a few dozen early chicks, no matter if we do prefer a use of the sitters later. The incubator is right here to stay and all of us who are alive to their possibilities are using them ... Get an incubator.'[22]

Many American producer/growers, like their Canadian counterparts, preferred to buy baby chicks from commercial hatcheries than run an incubator themselves. A report on the huge hatchery centre located at Zeeland, Michigan, in 1925 by the *American Poultry Journal* contained many comments by hatcherymen to the effect that poultry producers interested in hatchery-hatched chicks always bought day-old chicks and never incubated eggs themselves.[23] Even breeders often used commercial hatcheries to incubate their eggs.[24] Some went so far as to claim that the sitting hen as incubator was a thing of the past.[25] That trend had started by 1925, but hens still did most of the incubating. It was estimated that only 10 per cent of chicks came from incubators that year.[26] In 1930 about 42 per cent of American chicks were incubator-hatched, and by 1940 that number had increased to 73 per cent.[27] The difference in estimates of hen-hatched chicks from Graham's suggests he undervalued the percentage of hen-incubated chicks in Canada in the early 1930s. The rise in incubator-hatched chicks would be dramatic after 1940, however, in North America and parts of Europe. By the early 1950s about 98 per cent of chicks in Canada were hatched by incubator, 97 per cent in the United States, 99 per cent in Britain, 91 per cent in Italy, 99 per cent in Sweden, and 95 percent in Denmark. In contrast to these figures, 30 per cent of chicks in France were artificially hatched, 60 per cent in Austria, and 10 per cent in Greece. Considerable variation apparently existed in Europe with regard to hatchery-industry developments.[28]

At the 1916 meeting of the American Poultry Association, a group of hatcherymen got together and formed the International Baby Chick Association, which planned to promote the day-old-chick trade by advertising, and to campaign for uniform cheap shipping crates and shipment via government mail (which the association achieved by 1918). The hatchery industry continued to be quite unregulated in the quality of chicks it generated in the early years of the International Baby Chick Association's existence. The situation was difficult to control because the industry grew so rapidly in the 1920s. By 1918 there were 250 commercial hatcheries in the United States and by 1927 that number had risen to 10,000.[29] Volatility of production in relation to market demand was another problem indigenous to the early industry. It was often hard to stop oversupply of chicks, in spite of the fact that the hatching rates of eggs remained low. By 1921 only the best hatcheries could claim a hatching rate of 60 to 70 per cent.[30] Most were much lower. In Canada hatching levels stood at roughly 60 per cent in the late 1930s.[31] Even so, the huge capacity for output of American hatcheries often resulted in a glutted oversupply of chicks, which needed an outlet. Many chicks were simply dumped in nearby rivers, if they could not be sold at any price.[32] The volume that incubation technology made possible had profound effects on the entire industry. The sheer number of baby chicks which artificial incubators encouraged, for example, quickly shifted international trade patterns.

One major consequence of chick volume generated by incubator technology was a surge of American chicks on the Canadian market. Partially because of this American import problem, and partially because hatcherymen wanted an organization which could regulate quality in chick breeding, hatchery operators from Ontario and Quebec formed the Canadian Baby Chick Association in 1925 at a meeting in Guelph.[33] (In the late 1920s and over the 1930s various other hatcherymen associations, provincial and federal, developed to serve the hatchery industry.) The border problems with respect to baby chick sales and the rise of organizations designed to regulate chick standards reflected the fact that a specialized industry, devoted purely to hatching eggs and selling day-old chicks to producer/growers but not to breeding those chicks, existed as a result of advanced incubator technology. The growth of a hatchery industry that functioned separately from the breeding activity was encouraged by large rather than small incubators. In 1896 C.A. Cyphers in the United States invented an incubator that could hatch 20,000 eggs at a time. In 1908 the Mammoth incubator was patented.[34] Ads in the *American Poultry Journal* showed hatcherymen (and these were normally men) were selling 1000 day-old chicks a day by 1909.[35] These machines supported a business built on hatching eggs and also made it possible to reduce the cost of hatching substantially.

Poultrymen often dropped either breeding or producer/grower activities in favour of simply running a large hatchery. One American producer/grower, for example, started with a 100-egg incubator, with the intention of raising birds for meat sales. Quite quickly he began to concentrate on the day-old-chick trade rather than the poultry-meat market, because the market for chicks was so strong. By 1918 he had four incubators, and was selling the day-old chicks to his neighbours. His brother joined him in 1919 when they liquidated everything they had in their retail store in order to make room for a mammoth incubator. By 1921 they could not keep up with day-old-chick sales, and they also found it difficult to procure sufficient numbers of eggs for hatching from surrounding breeders. They undertook custom hatching too. By 1922 they sold day-old chicks to every state in the Union and to Mexico and Canada as well.[36] As early as 1911 these high-capacity incubators had started a new trend in the developing industry, namely, the move of businessmen who had never been poultrymen into hatchery affairs.[37] The fact that businessmen increasingly became hatcherymen encouraged the growing divide between breeder and producer/grower, that is to say, an increased separation of the breeder from the producer/grower. It was a situation matched by changes in other farming industries: namely, the involvement of business and the application of industrially oriented tactics to commercially focused farming.[38] Technology remained at the heart of the matter.

Hatcheries and Their Relationship to Breeding and Chick Production

The demand for hatching, not table eggs, rose dramatically because of incubator capacity.[39] The large hatcheries tended to rely on outside sources for hatching eggs, partially because their size and technology made them more labour intensive, but more particularly because increasingly the hatcherymen running them were businessmen, not poultrymen.[40] This reliance on outside sources for hatching eggs resulted in the rise of a new role for flock owners, although no one knew quite what to call people fulfilling that role. The word multiplier seemed appropriate, the *American Poultry Journal* thought. The selling of hatching eggs rather than table eggs was attractive to producer/growers because hatcheries paid 10 to 15 cents per dozen over the market price of table eggs.[41] Many producer/growers, therefore, stopped being producers and became multipliers. The way the multiplier/hatcheryman connection worked meant the hatchery provided the main contact between the breeders and producer/growers. The 'multiplier' multiplied the breeder's product for the hatchery, which in turn hatched the eggs and sold the resulting chicks to producer/growers. The hatchery only orchestrated how this exchange worked, but their decision on what breeder stock to perpetuate was critical to the way the breeder and producer/grower related to each other. It was ultimately the effect that hatcherymen had on what was accepted as superior breeding for multiplier use, and not whether incubation operators were breeders themselves, that became important to how egg-laying breeding proceeded from the late 1920s to the late 1940s. Hatcherymen directed breeding standards by deciding what breeding would supply the egg industry with producers, then subsequently telling multipliers what to 'multiply,' which in turn dictated what the producer/grower used to 'produce.' Hatcherymen's direction of breeding and subsequently of what multipliers would multiply made some contemporaries describe hatcherymen as either breeders or multipliers or both. While hatcherymen sometimes became multipliers, very rarely were they, as implied in the article quoted below from the *American Poultry Journal*, actually breeders.

> What is here referred to as multiplying is as yet not so well recognized as a distinct industry as are standard-breeding and egg-breeding ... But by whatever term they may be finally designated, they form the connecting link between the breeders on one hand and the hatcheryman on the other, and perform an important service in renewing and increasing flocks. Taking improved stock, which is the product of the skill of ... the egg-breeder, they allow the sexes to mingle for the purpose of producing hatching eggs for the hatcheries. They are not breeders in the most constructive meaning of the word, because they are dealing with large numbers and are not centering their attention on that careful selection and mating of individual birds which is the basis of all breeding progress ... Hatching is perhaps the most clearly perceived as a separate business ... Hatcherymen, and an increasing number of them, have gone beyond this stage of merely transforming eggs into chicks, and have become multipliers and many of them have become breeders.[42]

The Canadian Hatchery Industry, Record of Performance Breeding, and Government Standardization

The Canadian hatchery industry, partially in attempts to remain viable and maintain independence from the American industry, followed a common Canadian strategy: it chose to regulate and standardize itself through government agencies. The Canadian Baby Chick Association, shortly after its formation, focused attention on the hereditary quality of the Canadian egg-laying chicks its members sold and on the accreditation of the hatcheries themselves. If quality could be certified by an outside and more powerful authority, hatchery credibility would be enhanced, a situation that would encourage more sales, and the logical place to turn for such credibility was to ROP birds. Hatcherymen attempted to guarantee the breeding behind their egg-laying chicks for their customers by relying on government-endorsed breeding, thereby supporting the ROP breeders. Hatcheries bought ROP birds for their multipliers or persuaded multipliers themselves to buy ROP birds from breeders.

These efforts at directing the work of Canadian multipliers were made easier by the more general acceptance of the ROP, recording of egg laying, and egg-laying contests that existed in Canada, and by the dominating position the Canadian Baby Chick Association held in the hatchery industry. By 1928 the association controlled 60 per cent of hatching capacity in Ontario, for example, where one million chicks were produced every three weeks in the season (hatchery operations normally ran from late February until about June in this period).[43] The ROP was 'rapidly spreading its influence through the baby chick industry that has grown at an almost unbelievable rate in recent years,' the *Farmer's Advocate* explained. 'One farm alone in Ontario distributed more than 200,000 chicks this year, a great many of which were from Record of Performance stock ... Record of Performance blood [was] also being distributed through commercial hatcheries, using eggs secured from stock headed by Record of Performance males.'[44] Hatcherymen's support of the government-endorsed breeding structure of the ROP advanced the cause of entrenched systems, but did nothing to encourage any potential innovation that might emanate from genetic work being undertaken in the United States. Hatchery operations promoted the status quo when it came to breeding, and increased the prestige of many craft egg-laying breeders in Canada enormously.

W.R. Graham chaired a committee of the Canadian Baby Chick Association that intended to look into the possibility of an accreditation system which combined regulation of both hatcheries (as to the quality of their facilities) and the chicks themselves (as to the background of their breeding). In 1928 his plan was submitted to a meeting of the association, which approved it. The poultry division of the Department of

Agriculture in Ottawa promised cooperation. Basically the plan called for the following regulation:

> Every breeding bird [i.e., bird used to multiply] must be banded by a ROP inspector, signifying that the bird is of standard type, apart from minor disqualifications, and of good laying quality. All birds must be vigorous and healthy. No egg [can] be used for incubation that weighs less than 23 ounces to the dozen. The chicks shipped must be healthy and weigh not less than eight pounds per 100. The hatcheries will also be inspected for sanitation and hatchery records must always be open to the hatchery inspector as an assurance against fraud. It is estimated that the inspection under this system will cost the hatcherymen not more than five cents per bird banded.[45]

By 1929 the federal government and the Canadian Baby Chick Association had created the Hatchery Approval Policy, which combined regulation of hatchery production of chicks and 'certified' the breeding behind those chicks destined to serve the egg industry.

The American Hatchery Industry, Record of Performance Breeding, and Government Regulation

The heated growth of the American hatchery business in the early 1920s, if not international trade problems, had clearly promoted problems in the baby-chick trade in the United States. 'There are two common abuses in the matter of prices which have hurt the baby chick business,' the *American Poultry Journal* explained. One was advertising chicks at prices as low as 6 cents, while planning to sell only at higher prices. In fact, very few were offered at 6 cents. The other extreme was to advertise high-class exhibition and pedigree stock and fill the orders with poor chicks.[46] Both the International Baby Chick Association and the USDA were anxious by 1925 to correct this situation,[47] but the hatchery industry was hesitant to entertain any form of support for regulation that might appear to threaten its autonomy. M.A. Jull, senior poultryman with the USDA, called what he named the Standardization Conference at Kansas State College in Manhattan, Kansas, in order to draft a uniform plan to ensure the better national breeding and health of chicks produced by hatcheries. The draft became known as the Manhattan Plan. Jull described to breeders how the Manhattan Plan could lead to improvement of both breeding and, subsequently, the quality of eggs used for hatching in commercial incubators.[48] The regulations in various states differed from each other, and it was later decided that the word 'accredited' could not be used. Chicks were 'certified' under various state regulations, some of which applied to the breeding background of the birds, and some to inspection for health reasons.[49] The Manhattan Plan was not quickly adopted as a national one, partly because the International Baby Chick Association would not endorse it. Developments in the American hatchery industry reflected the stronger American (compared to Canadian) reluctance to involve government in business affairs.

Regardless, pressure mounted for better quality in chicks. Practical poultry keepers in the United States, the editor of the *American Poultry Journal* stated in 1925, wanted no part of breeding themselves, but they demanded quality chicks from hatcheries. American hatcherymen were forced to respond. 'The better hatcheries are bending every effort to produce chicks that will live up to such expectations,' the editor explained. 'The flocks that produce their hatching eggs are kept on farms, in the hands of farmers, and are farm poultry, not hot-house specimens. The males that are introduced to head up these flocks are production-bred cocks and cockerels that have vitality. They are carefully selected egg-bred flocks.'[50] While it made sense for American hatcherymen, as businessmen, to provide their customers with the sort of chicks they desired, they, like their Canadians counterparts, were not breeders. American hatcherymen, therefore, relied on the existing regulations concerning breeding put in place by poultry experts and breeders, namely, regulations orchestrated on a voluntary basis at the state level. State 'certification' of breeding quality for hatchery eggs revolved around use of an ROP, egg-laying contests, and adherence to the Standard of Perfection throughout the 1920s.

Evidence of American hatcherymen's support of the ROP could be found in the poultry press. In 1927 R.R. Hannas, a poultryman who wrote extensively in books and articles on many practical aspects of the chicken industry, reviewed the hatchery situation in Ohio for the *American Poultry Journal*. In his report Hannas stressed a growing concern of hatcherymen with chick quality by the late 1920s. He spoke of one hatchery owner, who, in his search for quality multiplier stock, turned to Canadian ROP birds. Mr Holzapple, owner of a hatchery in Ohio, purchased 'high quality, pedigreed stock from the ROP Association of British Columbia.' He bought one thousand pedigreed eggs, to be hatched, reared, and used in his multiplier flocks, Hannas told readers.

Holzapple was also a member of the Ohio Improvement Association and entered birds in the Ohio ROP, which started in 1926. He sold to producer/growers chicks 'certified' by an ROP background of the multiplier flocks.[51] Descriptions of another Ohio hatchery by Hannas in the *American Poultry Journal* confirmed how important the idea of beauty was to poultrymen concerned with good egg-producing ability.

> The Holgate Chick Hatchery, Holgate Ohio, is another hatchery where quality is distinctly noticeable. I saw [multiplier] flock after flock of Barred Plymouth Rocks of wonderful colour and barring, under the supervision of this hatchery and supplying hatching eggs to it. I have never seen so many Barred Rock flocks in such a short space of time that were such beauties. Mr. D.J. Groll, one of the proprietors, showed me cockerels in his barn that could easily have distinguished themselves at the Coliseum or any other show. Blood from such stock is put back into these farm flocks year after year.[52]

American hatcherymen may have been concerned about the ethical practices of chick sales and also about chick quality, but they had not been so sure about governments regulating their activities within the hatchery, and so they tried to resist accreditation of their operations. Certification of breeding was part of a movement that dovetailed with the more widespread emphasis on standardization which prevailed increasingly in agricultural circles early in the twentieth century. Pressure to adopt government certification for plant genetics had earlier dominated the seed industry,[53] and an ROP for poultry fit into that scenario. Accreditation with respect to how the industry itself might function, however, was something quite different to hatcherymen who saw themselves essentially as businessmen, not farmers or poultry experts, and therefore entrepreneurial in a fundamentally different way. The conjunction of business with farming interests resulted in clashing ideology when it came to plans for the accreditation of hatcheries. At the 1925 meeting of the International Baby Chick Association, the issue of 'accreditation' was discussed. M. Drumm, described by the *American Poultry Journal* as the 'patriarch of Missouri hatcherymen,' stated, somewhat ambiguously: 'If we are to put our business on the highest plane, we must have national certification, with terms having the same meaning in all states. But I oppose Government and State control. All the theorists in America want to get in with the Government. We in the hatchery business must use business sense ... We do not want the withering hand of Government paternalism to blight this business.'⁵⁴ In 1926 the president of the International Baby Chick Association, G.R. Spitzer, publically gave his support to accreditation of hatchery flocks, but was silent on hatchery accreditation.⁵⁵ The issue of hatchery accreditation was discussed again at the 1927 meeting of the International Baby Chick Association. The same tensions over control still existed. Hatcherymen might encourage better breeding quality in chicks, but government control over the hatcheries themselves was not very palatable.⁵⁶ By 1930 many believed the hatchery industry had started a trend to better-quality poultry found on the farms of either multipliers or producer/growers.⁵⁷

Along with these attempts to enforce the spread of better 'breeding' of hatchery egg-laying stock, the International Baby Chick Association tried to enforce more ethical behaviour among hatcherymen. Freedom to do business, it was now thought, might not be so attractive if even a few were not honest in their dealings. In 1932 a past president of the International Baby Chick Association, C.A. Norman, was accused by the association of unethical practices. Norman's response was to shoot and kill the current president of the association, C.B. Sawyer, and to wound the executive secretary, R.V. Hicks, before turning the gun on himself.⁵⁸

In 1931 a new federal plan for hatchery accreditation and ROP certification with more direct involvement of the USDA seemed nearer to becoming a reality in the United States. Some twenty-five states accepted the Manhattan Plan provisions and planned to approach the USDA.⁵⁹ But it was not until 1934 that a national program, overseen by the government, was put in place under an agreed-upon ROP structure. In 1935 the United States National Poultry Improvement Plan (NPIP) coordinated the project between the various states and the USDA, and ran the new United States Record of Performance certified by the federal government. Some poultry geneticists supported this new form of regulation. They and breeders both contributed to the formation of the plan, which, while much like the earlier USROP, involved more complicated recording than the earlier system. (By the 1950s the same individuals who had proposed it, rejected it, according to the Dutch geneticist Hagedoorn.)⁶⁰ ROP breeding in the United States now carried the same backing of government that it did in Canada, and thus extended the prestige of American ROP breeding. It proved to be significant for future patterns in poultry breeding, however, that some of the most important breeders (Prentice, Kimber, and the corn breeders who began experimenting with chicken breeding using geneticist aid, namely, DeKalb and Wallace) avoided the new federal ROP. It seemed to them that any form of an ROP overemphasized the role of the individual in selection strategies for breeding.[61] ROP breeding in the United States might have been added by hatchery support and federal endorsement, but, unlike in Canada, the most significant breeders continued to avoid such regulations.

The Hatchery Industry and Its Control of Chick Marketing: The Issue of Sexing Day-Old Chicks

Hatcheries might have orchestrated how stock passed to the producer/grower, but producers could exert pressure on hatcherymen. For example, producer/growers demanded different fare when technology made that possible and hatcheries would be forced to incur additional costs in order to meet these demands. Producer/growers wanted a better guarantee of egg-laying pullet chicks. The emphasis on egg production made female chicks more desirable than male, but it was virtually impossible to distinguish the sex of day-old chicks by simply looking at them. In Britain, as early as 1919, R.C. Punnett had begun work on chick sexing by feather colour. In 1923 he reported sex-linked feather-colour characteristics. Punnett then went on to establish a way to sex-link feather colouring to bird breeds used by commercial egg producers, thereby commercializing his work.[62] Chick sexing was soon commonly practised on the basis of the colour feathering method in Britain, where farmers, intent on egg production as were their North American counterparts, preferred pullets over cockerels. The extra charge the farmer paid for the resulting pullets made the sexing economical from the hatcheryman's point of view. From the farmer's point of view, he saved the expense of raising 50 per cent of chicks that he did not want.[63] The desire for pullets was just as strong in North America because of a similar focus on egg production, but the Punnet method was not extensively used there.[64]

It was the advent of a different type of chick sexing that worried North American hatcherymen. By 1933 Japanese experts had devised a reliable way of sexing day-old chicks which did not rely on sex-linked feather colour. With the possibility of sexing the White Leghorn, which cannot be done on the basis of feather colour, chick sexing by the new method became attractive to North American table-egg growers using Leghorns, particularly in the western regions of the continent.

The move to sexing initially took hold in British Columbia, where the agricultural college at the University of British Columbia introduced it, and subsequently gathered momentum in American Pacific states, where the Leghorn dominated the poultry industry (and where also the chicken meat industry was less significant). Highly trained Japanese experts could sex chicks quickly and accurately by palpating their vent (sex) area. One person working in British Columbia, for example, sexed 10,000 chicks a day with an accuracy of 98 per cent. American poultry experts feared an influx of Japanese immigrants who would establish chick-sexing companies. Americans and Canadians quickly established schools for chick sexing to forestall that situation. R.M. Forsyth, for example, came from British Columbia to Ontario to teach sexing techniques. Since the greatest demand was for pullets, due to the nature of the poultry industry, which focused primarily on egg production, eschewing the sex of chicks for sale introduced a number of issues for hatcherymen – fear of excess pullets, a glut of table eggs, and insufficient numbers of cockerels – which in the end would hurt the meat/broiler side of the industry, a sector of more importance to eastern, than to western, poultrymen.

Eastern hatcherymen in both the United States and Canada bowed to the inevitable.[65] Sexing chicks was expensive and time consuming, but their trade in baby chicks with western regions necessitated the move. Soon producer/growers in the eastern United States (chiefly New England) and Canada (mainly Ontario) demanded better assurance that they were purchasing only pullet chicks. Comments such as the following on the Ontario situation make it clear that sexing was forced on the hatchery industry because producer/growers increasingly wanted sexed chicks. 'Hatcherymen in Eastern Canada are not preparing for chick sexing because they welcome the practice or consider it a good thing for their business or the poultry industry ... Ontario hatcherymen have not invited chick sexing. They are just accepting it as the inevitable.' By 1935 all Ontario hatcheries offered day-old pullets. One important Ontario hatcheryman, F. Bray, went so far as to advertise the fact that he had hired a man who had been trained by Japanese experts in British Columbia to sex chicks.[66] With the advent of chick sexing, eastern hatcheries soon found that indeed they frequently could not market the cockerels. Hatcherymen began to offer cockerels to farmers or their wives, free of charge, on a barter basis. Some hatcheries provided free feed to their producer/grower clients too.[67] (By this time feed companies had allied themselves with hatcheries, by holding conferences on hatchery matters and on how the multipliers and producers/growers functioned.)

As a matter of interest, the sexing of day-old chicks has continued to be an aspect of the chicken industry. Chicks destined for the table-egg and meat industries are sexed. In the case of breeds which demonstrate sex-linked feather colouring, sorting – generally using the dominant silver and recessive gold gene, but also the dominant barring or striping and the recessive lack of stripes – by feather colour became common. In the case of the Leghorn, sexing by feathering can now be done, but it is the rate of feathering as the young birds grow, not the colour of the feathers that makes this possible. Fast-feathering males (young birds that feather comparatively early) mated to slow feathering females (young birds that feather relative late) will yield fast-feathering pullets and slow-feathering cockerels. Distinguishing sexes at the hatchery can easily be done by either of these methods and does not require the labour that vent sexing under the Japanese method did.[68]

The Hatchery Industry and Its Control of Chick Marketing: The Issue of Started Pullets

By the 1930s hatcherymen in Canada and the United States found themselves reeling from a second producer/grower demand: namely, the desire for started pullets (meaning birds that were not simply day-old chicks).[69] Started chicks came from internal hatchery issues and a different use of technology. Chicks over forty-eight hours old, if not sold, had to be dealt with. Overproduction brought on the problem. If chicks were not destroyed, they had to be fed. Started chicks had attracted producers in the United States as early as the 1920s.[70] These chicks were seen increasingly as a problem after 1936 in Canada.[71] Rising demand for started chicks, however, forced the hatcheries to supply that product to producer/growers.

The Hatchery Industry and Its Control of Chick Marketing: The Entrance of Genetics Combined with Corporate Enterprise

By the early 1940s hatcherymen faced a different and particularly serious issue that threatened their independence and ultimately the very existence of many of their operations. Conflicts within the American Poultry Association, widespread dissatisfaction with egg-laying recording and contests – and therefore with an organization like the

ROP – the increased division between breeder and producer that the hatchery industry encouraged, chick sexing, and the started-chick problem coincided with important new developments in relation to genetics: namely, the entrance of corporate involvement. By the late 1930s corporate enterprise in the United States had begun a pattern which in the end revolutionized breeding for the table-egg industry. Initially, it was the success of the hybrid-corn heterosis breeding method as applied to plant breeding that made American companies want to explore the idea of using the same system for the production of egg-laying chickens. Unlike many scientists working at research stations in the 1930s, the managers of the corn companies believed the hybrid-corn breeding system could be successfully applied to chickens. They were willing to finance expensive experimental programs in order to find a way to achieve this end. Some corn companies, which already used professional geneticists, began hiring scientists to undertake poultry breeding operations designed to produce what was known as 'hybrid' chicks, namely, birds that had been generated by some form of breeding involving hybridizing. The Wallace family – that is, Henry A. Wallace, who developed the hybrid-seed company Hi-Bred Corn Company in 1926 (renamed Pioneer Hi-Bred Corn Company in 1935), and his son Henry B. Wallace – initiated this effort to produce commercial hybrid chicks in 1936. By 1942 the Wallace family was selling hybrid egg-laying Leghorns under the name of Hy-Line.[72] It is important to note, however, that selection methods used to achieve these hybrids chicks were generally kept secret, even if inbreeding and line crossing lay at the heart of the matter. So, of course, was the genetics being used for the breeding programs, but all would have belonged to different Leghorn strains. The hybrid-corn breeding method was generally applied to poultry within one breed, and most frequently to the egg-laying specialist, the Leghorn. There were a variety of ways for geneticists to produce the four inbred pure lines that went ultimately into the final terminal cross. One could set out to establish many inbred pure lines, designed to breed truly, by the closest inbreeding possible (brother to sister), and through time let the ones that amassed significant lethal recessives to die out naturally. Everyone knew that certain lines tolerated inbreeding better than others, but only through actual breeding could one know for sure which ones they were.

Another, perhaps more productive way, was to carry fewer inbred lines under perhaps less intense breeding practices, using Wright's path coefficient, in order to produce more individuals within those lines. This situation allowed the development of sub-lines – thus making possible a more sophisticated process of selection for birds to be used in future intense inbreeding programs. Otherwise, the geneticist was forced to rely on a line chosen simply because it could tolerate intense inbreeding. The true-breeding lines that were selected for the heterosis cross should be known to nick well with each other as well. Generally speaking, whatever method a geneticist chose to produce the four initial lines, breeding then worked as follows. Inbred line A was produced by three generations of inbreeding of brother to sister. Line B was created the same way. Inbred line A and B were mated. Resulting sons of the match were kept, the daughters discarded. Line C and line D were produced the same way as line A and B. When line C and D were crossed, only the daughters were kept. Then AB sons and CD daughters mated to produce the commercial chick.[73] From the beginning to the end of any such program, large numbers of birds are needed in order to quantify varying results.[74] The hybrid-corn breeding method relied on huge numbers of birds. The method for poultry, then, was immensely expensive. Only large breeding establishments could undertake such a procedure. Even government agencies, let alone an individual farmer, could never have incurred such cost. The hybrid-corn breeding method as applied to chickens, then, was designed for corporate work.

By the early 1940s American geneticists working with breeding companies were in a position to offer egg producers a hybrid hen with increased egg-laying capacity on a consistent basis via the hybrid-corn breeding method, and American breeder companies of egg-laying birds began to franchise hatcheries, which were themselves independent, in both the United States and Canada. Hatcheries advertised what they had been franchised to sell.[75] The work and fortunes of Ezra Neuhauser serves as an example of how this worked for American and Canadian hatcheries. Born on a farm in Indiana, Neuhauser raised purebred birds from the time he was seven. When he was twenty-three he was operating a hatchery in Ohio, and by the 1930s his business had expanded to such a degree that he ran hatchery operations in a number of locations. He sold chicks into Canada too. A Chatham veterinarian in Ontario approached Neuhauser about the possibility of hatching chicks for him in Ontario, and by 1939 Neuhauser had a number of hatcheries in Ontario, operating under the name of Neuhauser Hatcheries Ltd. In the 1940s he was selected by the Hy-Line breeding company to franchise the Hy-Line layer to various hatcheries throughout North America, thereby gaining widespread control over the distribution of this early hybrid table-egg-laying hen.[76] Any hatchery in Ontario that could act as a distributor of Neuhauser's Hy-Line chicks was likely to survive.[77]

Canadian hatcherymen in many Canadian provinces were forced to confront the fact that producer/growers liked egg-laying birds resulting from the crossing of breeds or of strains within a breed, even if these emanated out of the United States.[78] ROP birds were not wanted, and few if any hybrids were available from Canadian breeders. At the 1941 meeting of the Ontario Hatchery Approval Association there was heated discussion about hybrid chicks. One hatcheryman who ran a large operation with incubators in a number of locations, A. Seiling, stated that 65 per cent of his sales that year had been hybrids, and he prophesied that would increase the next year. He considered it ridiculous that birds used to produce hybrid chicks could be disqualified for commercial-breeding purposes because of minor issues relating to the Standard. He added that once a producer had experienced hybrids it was impossible to sell him/her anything else.[79] That same year E.S. Snyder of the Ontario Agricultural College discussed hybrids and the Canadian breeding of them in an article in the *Farmer's Advocate* entitled 'The Pros and Cons of Hybrid Chickens.' Snyder noted that, whereas in the United States the vast majority of birds raised for meat purpose were cross-breds, an emphasis of the rearing of chickens for meat in Ontario was still not profitable enough to warrant such specialization. Whether cross-breeding for egg production was a sensible plan, he did not know from first-hand knowledge. 'Few authentic comparisons of the egg production of cross-bred progenies and standard bred progenies have been done in Ontario,' he stated.[80]

Both Hy-Line and the DeKalb Hybrid Corn Company, American organizations, marketed chicks through franchised Canadian hatcheries in the 1950s.[81] By this time DeKalb was maintaining over 250,000 birds in its breeding operations and sold under franchise all over the world. In 1944 DeKalb had hired J.H. Martin from Purdue University to set up their poultry project.[82]

Canadian hatcheries that focused on the selling of egg-laying chicks were saved from going out of business by franchising, because, increasingly, adequate volume with good uniformity was vital – something that the large companies could provide in a way that smaller breeders could not. Those that survived tended to be attached, via the franchise system, to a breeding company. It was largely American breeders who now held control over the future of the Canadian hatcheries, by virtue of the fact that these companies decided which hatcheries to enfranchise. Hatcheries were no longer in a position to dictate to breeders by deciding what breeders they would support. Breeders decided which hatcheries would continue to exist via their support. Under these conditions, hatcheries often also acted as multipliers. The *Farmer's Advocate* explained: 'Most [hatcheries] are big and most of them are "franchised" establishments operating under contract with a chick breeding company. The breeding company supplies the foundation eggs from which the hatchery operator will produce his own breeding flock. This breeding flock will supply him with his commercial eggs.'[83] While the franchise system weakened the position, relatively speaking, of hatcheries within the poultry industry, the days of their very independence were also numbered in both the United States and Canada. The feed companies became involved directly in the egg industry by contracting with the producer/grower, buying franchised hatcheries, and packing and marketing eggs.[84]

Decline of Record of Performance Breeding in Canada and the United States

The rise of the breeder companies clearly spelled the death knell of the ROP in Canada. Many Canadian poultry experts who had adamantly supported the ROP/standardbred structure had come to believe that the standardbred bird offered the commercial egg farmer nothing.[85] The day of the 'purebred' or 'standardbred' bird as producer was drawing to a close, but not without considerable resistance to the idea of hybrids or cross-breds in some provinces. Two refused, not only to accept such chicks in hatchery supply flocks, but also the movement of hybrids or cross-breds from other areas across their border. Fear of 'mongrelizing' remained as strong as faith in the ROP in these locations throughout the 1940s.[86] Meanwhile, the shift elsewhere to hybrids or cross-breds was rapid. In 1951, for example, some 48.6 per cent of chicks hatched in approved hatcheries in Ontario were cross-bred, compared with 24 per cent in 1927.[87] Most of these would have been American egg-laying hybrids franchised to Ontario hatcheries. Even the pure Barred Rock was on the way out in Canada.[88] Canadian ROP breeders, especially in Ontario, found that fewer and fewer egg producers/growers wanted to buy ROP stock. That situation, of course, meant that Ontario hatcheries did not encourage multipliers to use ROP stock. ROP breeders could get very little money for their birds, and they predicted that this trend to poor remuneration would bring about the ruination of the Canadian table-egg industry. Some ROP breeders argued that the regulations under the Canadian program should be more elastic, and suggested that 'each ROP breeder be free to conduct his or her breeding program, which appear[ed] most desirable for the continued improvement of all economic characters in the breeding stock.'[89]

ROP breeding in the United States, never as entrenched as in Canada, was also on the way out. British experts noted the frustration of American ROP breeders, who chafed under the restrictions of the system's provisos, and believed that genetics had managed to penetrate the breeding ranks of traditional breeders of egg-laying stock because of the system's rigidity.[90] American breeders, they noted,

> feel that they now want to be free of the shackles of the breeding provisions of ROP; they want to be free to adopt any of the many new ideas of poultry-breeding ... The college geneticists in the US, many of whom have always been skeptical of the breeding provisions of the ROP, welcome this revolution. They argue that at the Colleges they have not the facilities or numbers to indulge in satisfactory work in poultry population genetics. Under the new set-up the practical progressive breeder can launch out on many new ideas, unshackled by ROP, the college geneticist can cooperate, advise and study the results objectively, knowing that the progress or lack of progress made by such breeders will be accurately assessed ... In England we are fortunate in that we have not started official registration of poultry.[91]

These words indicate that direct, geneticist involvement in the breeding industry appeared to play as much a role in attacks on the ROP as ROP-breeder dissatisfaction with the system. The passage also shows how strong the conflict in the United States was between traditional attitudes towards breeding, by that time seen as the ROP, and geneticist views. In light of what was going on in the poultry-breeding world of the United States, the British experts believed it would be foolish to start an ROP process in England in the 1950s.[92]

The entrance of corporate enterprise in the United States into the egg chicken-breeding world clearly had a profound effect on the dynamics of the hatchery industry, and on how both chicken breeding and the egg industry evolved.

The Effects of Corporate Breeding on Genetic Education and Research

Corporate breeding also initiated a change in the way genetic research applied to chicken breeding would reach the American public. Traditionally, the dominant centres for poultry breeding research had been found at agricultural colleges, experiment stations, government agencies like the Bureau of Animal Industry under the USDA, and universities where aspects of evolution theory, chromosome theory, and so on were studied. Even if much of this work done on poultry before 1930 was of little use to farmers, the philosophy behind public support for such undertakings initiated at agricultural colleges and experiment stations arose from the conviction that American farmers were the expert breeders, and any information available should be aimed at them in order to improve their breeding methods for egg-producing hens. After 1915, if not before, geneticists clearly hoped ultimately to reach breeding farmers directly through the research they did at government-funded agencies.[93] Increasingly by the late 1940s, though, geneticists produced information of more interest to the developing new private companies, which then sold the resulting product to farmers who earlier might have been breeders, or to farmers' wives who had always been producer/growers. Breeder companies did not disseminate information to these people on how to breed. They were perceived to be consumers of expertise practised by someone else, not learners of it.

Under these conditions, American university geneticists served the interests of the companies and not farmers. Increasingly, men like Lush acted as advisers to geneticists working in companies, or else trained people who eventually would work for these organizations. By 1950 the demand for trained geneticists by the breeder companies was so great that it outstripped supply. With fewer breeders and the consolidation of breeder companies into smaller numbers by the mid-1960s, the need for trained geneticists declined.[94] Industry growth, which advanced in spite of company contraction, went hand in hand with the decline of many academic programs. Some forty-five poultry departments were active in the United States at the peak of poultry education in the 1940s, only to disappear rapidly in the 1960s.[95] The tendency of government-supported institutions in the United States to spearhead independent poultry-breeding programs for improvement through genetics became less and less prevalent. Competition between breeder corporations demanded that they better rationalize production. As a result, they began to take over much of the research role that American universities and experiment stations had traditionally performed.[96] These trends echoed those in plant-breeding industries, particularly corn, where companies took over the work of agricultural colleges and research done by such organizations as the Bureau of Plant Industry.[97] The move to company-dominated experimenting programs concerned some university geneticists, who feared it brought with it the dangers of self-serving research, and therefore research that did not advance pure science. 'What breeder,' F.B. Hutt asked in the 1960s, 'having discovered something that gives him an advantage over his competitors, will be altruistic enough to tell [others] about it?' 'Furthermore,' Hutt added, 'there are many kinds of research in which breeders have no interest whatever, and [other kinds that] might almost put them out of business.'[98] Many others working at agricultural colleges and experiment stations noted the growing trend in the 1970s with dismay.[99] The decline of poultry-related research at academic institutions that occurred across the United States after the 1960s resulted from the impact of complicated issues,[100] but the move of breeder companies into research to serve their needs seems to have played a role in the trend. (Somewhat ironically, by the late 1990s, as a result of the shrinking attention to poultry science in academic institutions, the companies found they confronted a shortage of people trained scientifically in poultry matters.)[101]

The theoretical concept that academic research in breeding was undertaken in order to teach farmers how to breed stayed in place much longer in Canada, perhaps partially because the quantitative approaches of agricultural genetics did not affect breeding research in that country until after it had done so in the United States. Information from genetic research had flowed into Canada from Pearl's time, but Canadian research institutions had not, throughout the 1930s, used geneticists to the degree that Americans had to test better breeding strategies for egg-laying chickens, especially those that involved extensive inbreeding of lines and outcrossing. No private individual undertook the expensive experimentation that Henry Wallace had, and the government-funded agricultural colleges were slow in even attempting to do so. Quantitative genetics and Lush principles did not play a role in Canadian breeding of any livestock until the late 1940s; that is, after the genetic/craft conflict had been largely resolved in the United States and corporate (not state) hybrid breeding had come to dominate at least the chicken-breeding scene. Research on heterosis in chickens began, for example, at the Ontario Agricultural College in 1945 by inbreeding strains of Barred Rocks, but this work was not undertaken by scientists.[102] The college soon realized that untrained staff could not properly undertake complicated quantified cross-breeding programs. F.N. Jerome, who had been a student of Graham's and gone on to obtain a PhD in genetics, joined the poultry department in 1948.

Scientifically trained men were not prominent in Agriculture Canada's chicken affairs either before the late 1940s. F.C. Elford and G. Robertson, the directors of the poultry division until 1946, had both been practical poultrymen with no formal education in science. Quantitative genetic principles were not unknown to the research branch by the 1940s (J. Stothart had introduced them, at least theoretically, as early as 1934); but experiments in breeding, run by Agriculture Canada at various government experimental stations, did not incorporate them. ROP Leghorns from various breeders were collected and subsequently used in attempts to find better producing strains that bred truly. Inbreeding was deliberately avoided until 1947. By 1949 it was apparent that this form of 'pedigree breeding' (as the department called it) had not increased egg laying in the birds over the past ten years. The research branch of Agriculture Canada decided to change breeding strategies.[103] In 1950 the department hired an agricultural geneticist, R.S. Gowe, who introduced quantitative breeding programs to experimental work involving chickens undertaken by the Canadian government. Gowe worked with a number of Leghorn strains between 1950 and 1973 for Agriculture Canada, in order to supply improved strains to the Canadian breeding public. He also investigated genotype–environment interactions in poultry. While he managed to breed improved Leghorn strains which were made available to the Canadian breeding public, perhaps his most important contribution to poultry breeding was the establishment of a line against which others could be tested in order to assess the role of environmental factors versus genetics in egg production.[104]

The idea that breeding research should help farmers continued to drive experimental chicken-breeding activities in Canada even after quantitative genetics was incorporated into such research. This philosophy dominated the thinking of geneticists at agricultural colleges as well. In spite of the move to better-trained scientists for public research, though, by the late 1960s it was apparent that the experimental efforts of Canadian institutions were not changing farm breeding in Canada. Breeding companies, mostly American, began to control the Canadian poultry-breeding scene in the 1950s, and although academic breeding research in Canada continued to be designed, in spite of the infiltration, to help farmers breed better, that situation would not last.

For one thing, by the 1960s it was no longer easy to acquire good foundation poultry material, which had resulted from the new breeding principles, to work within research experiments conducted at Canadian institutions. Developments in the United States, namely, the involvement of companies, had altered how the market for chicken genetics operated. The breeder organizations still might sell breeding material to other breeders, but they maintained a biological lock on their breeding by offering only one sex for sale. The Canadian government managed to overcome this problem to some degree, and brought a number of the best combining American strains into Canada for the use of all Canadian breeders. The Ontario Agricultural College was able to buy Mount Hope Leghorns. The crossing of these lines on another that the college developed resulted in superior egg-laying strains being available by 1960 to the Canadian public.[105] Most Canadian breeders did not flock to the new Canadian strain-crossed birds. Die-hard supporters of ROP philosophy refused to work with any change and certainly not with the new genetic principles, that is, methods revolving around complicated three- and four-way crosses. Even after 1945, when the Canadian government funded the work of a geneticist, S. Monroe, to experiment with the idea of inbred lines that were designed for crossing and would work within the ROP system, most breeders rejected both the lines and the new quantitative genetic techniques that had been applied to them. Gowe (who became director of the Animal Research Centre in the Research Branch of Agriculture Canada from 1965 to 1986) found in the 1950s and 1960s that he could get very few breeders to understand the new breeding methods practised on poultry that emanated out of the United States.[106] The failure to address what genetic breeding meant would damage Canadian breeders' ability to compete in an increasingly international world.[107] It was a situation that also encouraged the termination of government-supported breeding programs designed to help private breeders. This move away from public-sphere breeding led ultimately to the downfall, not just of general farmer breeding of poultry, but also of breeding independence in Canada. American breeder companies became the main suppliers of poultry genetics in Canada through hatcheries now controlled either by them or by what was known as integrator-companies (discussed in the next chapter).

The Role of the Hatchery Industry in the Cleavage between Breeders and Producers

The North American hatchery industry had introduced a subtle change in the relationship between the breeder and producer/grower as early as 1911. The growing pattern of buying day-old chicks from off the farm and from non-breeding hatcheries reduced the involvement of producer/growers with breeding before the advent of commercial hatcheries. They had traditionally acquired stock from breeders and subsequently made decisions on how to multiply or reproduce those genetics. The move to relying on hatcheries, not breeders, for birds started a trend that would be on-going: namely, the 'deskilling' of producer/growers in breeding matters and the reduction of their control over the genetics they would use. The producer/grower might have forced the hatcheries to return to cross-bred stock by the late 1940s, and thereby helped bring about the downfall of a hatchery philosophy that favoured the breeding of egg layers; but in the end, producer/growers would have no say in how poultry were bred, nor any choice in what stock they would use. When breeder companies replaced hatcheries as directors of breeding operations, producer/growers became even less independent as to what genetic material they would use.

It was the hatchery industry in Canada and the United States that orchestrated the rise of a more widespread uniform, and government-regulated, poultry-breeding structure for the egg industry. Improved breeding for egg layers meant ROB breeding and was endorsed as such by hatcheries and government. The involvement of the hatcheries this way in breeding encouraged the already existing tendency for ROB structures to appear as breeding methodology. Actual methodology used to produce egg-laying poultry became even more hidden. The shift to the use of corporate-regulated breeding changed the position of the hatchery industry in poultry affairs, and the producer/grower aided in that transformation. The hatchery industry was consolidated into fewer and larger operations as franchising patterns spread. Hatchery independence as a business concern began concurrently to erode. In 1951, for example, there were about 450 hatcheries in Ontario, with a capacity for 14 million eggs. In 1961, 225 hatcheries in Ontario were putting out the same volume – 14 million.[108] There was another factor, however, in the downfall of the ROP. The ROP had been designed for the egg industry, and when a new emphasis on poultry meat arose in the 1950s, particularly in the United States, many breeders no longer focused on eggs, but rather devoted themselves to meat breeding. That factor alone reduced the hegemony of the ROP as a breeding method, for reasons to be discussed in the following chapter. Rising concerns with meat breeding in the end cemented a complete poultry-breeding revolution. The story of the meat industry, then, is the story of the culmination of geneticist hegemony over craft breeders. The meat or broiler industry is the subject of the next chapter.

Chapter Six

North American Chicken Breeding and the Rise of the Broiler Industry

From the beginnings of a commercial industry in the mid-nineteenth century, attempts to increase egg yields dominated poultry-breeding strategies in North America. All major, craft, or traditional breeding schemes from the late nineteenth century into the twentieth were aimed at creating superior egg-producing hens. When American geneticists addressed poultry breeding early in the twentieth century, they too were interested in egg productivity. Corporate involvement in the breeding of chickens, emanating out of the United States, also focused on increasing the number of eggs that hens laid. The story of North American chicken breeding before the mid-twentieth century, then, is primarily one of the egg-laying ability of birds. Breeding for poultry meat did not receive the same attention from either craft breeders or scientists. The meat industry was perceived to be of secondary importance to the egg industry. By the late 1930s, however, both craft and American geneticist breeders started to make more concerted efforts to find ways of improving meat production. By the 1950s geneticists had fundamentally taken over the job of breeding for meat within a corporate structure, and simultaneously a revolution in the business organization of the broiler or meat evolved. Chicken growing quickly became a global industry, primarily through the American meat industry's new structure and corporate breeding strategies.

In this chapter I focus on breeding and the chicken meat (or broiler) industry in North America. I assess what changed with respect to breeding specifically for meat, and outline how the business organization of the broiler industry concurrently was revolutionized. The American poultry meat story has dichotomous features. In many ways American breeding strategies, used in meat production, reverted back to eighteenth- and nineteenth-century breeding theory, while concurrently the poultry business world experienced a form of twentieth-century corporate integration. The new poultry-meat industry, as it evolved in North America, rested primarily on broiler (young cockerels), more than on capon (castrated adult male) production. The word 'broiler' seems to have originated as an American term for cockerels slaughtered for meat at a young age, as opposed to older castrated males.[1]

Meat Production and Marketing in the Early Twentieth Century

A great deal of poultry meat generated on farms until at least the end of the nineteenth century was intended for home consumption or for local barter markets. When poultry meat became remotely commercially viable at the beginning of the twentieth century, it simply worked in a subsidiary fashion with the egg industry. Meat, whether generated for home consumption, barter trade, or the commercial market, tended to be derived from either surplus cockerels or older egg-laying hens, described as 'spent' hens, meaning birds that had finished their egg-laying careers. This trend was particularly dominant in Canada. As the Ontario government explained in 1916:

> The question of poultry production readily divides itself into two branches which, though in a way separate and distinct from each other, are yet very closely linked as we find the poultry business carried on in this country. The production of eggs is, undoubtedly, the prime object of the vast majority of those engaged in poultry keeping, and is without doubt the most profitable branch of the poultry business. The production of meat is and ever shall remain a secondary branch of the work. With practically all our poultry products coming from the farms, where poultry is kept largely as a side line, the only meat on the market is surplus cockerels and cull pullets sorted from the young stock grown to renew the flock of layers.[2]

The avenues for selling poultry meat into a commercialized market in North America from the late nineteenth century to the early twentieth were as complicated as those for eggs. The poultry producer might contract birds to an end customer, such as a hotel. Often producers sold their birds live to either the processor, or a fattening plant. The biggest markets, however, remained large urban centres, and poultry was shipped to the final consumer through a complicated middleman chain which took either live or dead birds. Before 1930 poultry meat most commonly came on the market in the fall, namely, after the egg-laying season was over. New York became a focal urban point for live chickens due to its heavy Jewish population. Live poultry could be killed according to Jewish traditions. It was estimated that by 1926, about 80 per cent of the poultry consumed in New York were bought by Jews.[3] (Receipts on markets of live birds began to decline in the late 1920s in the United States, although the selling of live poultry in terminal markets persisted in a limited way until the 1950s).[4] Dead meat was processed in similar fashion in both the United States and Canada. The birds were not eviscerated before the 1940s, nor were their heads and legs removed. Processors plucked the feathers and drained the blood. The brain was often removed as well. Poultry meat presented on the market in this fashion was known as 'New York Dressed.'[5] In 1942 the United States Army Quartermaster demanded that birds be eviscerated (and the Canadian government urged the processing plants in Canada to comply), and 'New York Dressed' became a thing of the past by the 1950s.

The emphasis on egg production did not mean that income from poultry meat played an insignificant role in the poultry economy in the early twentieth century, especially in certain geographic areas. Data from a study done in Connecticut indicated that by 1924 almost half the poultry product revenue was generated by meat – some 43 per cent versus 57 per cent for eggs.[6]

Producer and Breeder Attitudes to Specialization Breeding for Egg or Meat Purposes in Chickens

In view of the enormous emphasis on egg production in the North American chicken industry and the reality that meat also played a large part, it is interesting to note that before 1920, poultry keepers, who were generally women, preferred to work with dual-purpose breeds or their crosses rather than the single-purpose egg breeds for table-egg production. The better meat compensation these larger breeds provided, over that generated by fowl from a table-egg specialist like the smaller Leghorn, made poultry keepers find the dual-purpose birds more remunerative on an overall basis. There is evidence of this dual-purpose breed preference in the United States from slaughter figures. Since all birds ultimately went for meat, data on stock coming on the market reveals something about the general poultry population used for either meat or eggs. In 1924 it was calculated that 50 per cent of American birds destined for slaughter came from dual-purpose lines, 25 per cent were egg-specialized Leghorns, and only 25 per cent were from unknown background.[7]

'The dual purpose fowl is really more numerous and in a total of its various breeds is more popular than either the specialized egg fowl or meat fowl ... The fact remains that the dual purpose type is most popular where the balance between eggs and meat is maintained,' the *American Poultry Journal* stated in 1922.[8] And balance was what most poultry keepers wanted, although the journal did not seem to condone that reality. The breeding of birds specifically for meat purposes had commanded no attention for some time, the *American Poultry Journal* reminded its readers in 1927, saying that 'chicken meat is largely a by-product of egg production ... For twenty-five years or more ... there has been a growing group of people who devote their attention to improving the laying qualities of chickens.'[9]

The purpose focus did not differ in Canada. Testimonials in the farm press reveal widespread support for dual-purpose lines.[10] The dual-purpose Rocks or their crosses were most favoured for both egg and meat production at the beginning of the century, as the census for Canada of 1901 revealed. The ratio of pure Rocks to Leghorns was close to three to one.[11] The *Farming World and Canadian Farm and Home* survey of 1903 on farm poultry indicated how significant the genetics of the dual-purpose breeds were to the commercial industry. Some 57 per cent of flocks surveyed showed the influence of dual-purpose breeding, while 23 per cent came from egg-specialist birds.[12] A more detailed study done in 1908 on farm poultry in one Ontario county revealed much of the same dual-purpose emphasis: of the 15.5 per cent of purebreds in this county, 11.8 per cent were Barred Plymouth Rocks and 1 per cent Leghorns. While cross-bred or grade stock clearly dominated the poultry population of this county at 84 per cent, some 46 per cent were predominantly Barred Plymouth Rocks and 7 per cent were mainly Leghorns.[13] By 1921 there seemed to be a move towards Leghorns – still used, however, for both meat and egg purposes. While Leghorns had come by 1921 to outnumber the Rocks in Canada, when combined, the two breeds constituted 64 per cent of the nation's pure stock.[14]

Specialization division came last in poultry breeding compared to the other North American livestock industries. Selecting cattle for dairy characteristics, while significant to varying degrees in parts of northern Europe for centuries, had not been important on a large scale in North America until the beginning of the twentieth century, but that was still years ahead of attention given to separating egg from meat use in poultry, a situation not fully in place until nearly 1950. Emphasis on milk production in dairy-cattle breeding would reach new heights in North America, compared to Europe, by 1920.[15] Beef emphasis in cattle had dominated the breeding of stock from Bakewell's attempts to increase the meatiness of Longhorns to the hegemony of the Scotch beef Shorthorn in the cattle population. Beef cattle were the major arm of purebred cattle in both Canada and the United States from the early nineteenth century until into the twentieth.[16] Heavy or light styles in horses had been sought for nearly 100 years in Britain and North America.[17] Sheep had been bred for wool for 600 years in Europe (in some cases much longer, as early as the pre-Christian era in Spain), and for meat for 400 years in Britain.[18] It is a curious paradox that the livestock species today most effectively influenced by genetic breeding strategies, poultry, would not undergo a fundamental division in emphasis in North America until long after single-purpose breeding was practised on other types of livestock capable of yielding different products. While complicated issues concerning the capacity for a poultry-meat or broiler industry to exist had to be solved before separation of purpose made sense, the position of women within the poultry industry seemed to be part of the story.

Concerted efforts at divided-purpose breeding did not appear historically in North American livestock industries when female hegemony in commodity production prevailed: that meant dairy cattle arguably before 1920 and all poultry before at least the late 1940s. Livestock industries, where production was traditionally controlled by men, saw an emphasis on single-purpose breeding much earlier. Sheep, beef cattle, and horse husbandry, as well as breeding, had always been male occupations in both Britain and North America. Whereas beef-cattle breeding and feeding was generally done on farms by men, dairying might be undertaken on the same farms by women. An American beef-cattle farm of the 1880s, for example, also maintained milking cows under the supervision of the farm wife.[19] The greater emphasis on the breeding of cattle for milk specialization in a more widespread way did not begin in earnest in North America until the hegemony of women had effectively been removed from dairying, and at the same time that sectors of the industry took on a more important commercial nature.[20] The shift to factory cheese in the 1880s in Canada, for example, had not been enough to end women's important role in the industry. The rise of creamery butter and a viable fluid milk trade undermined the position of Canadian women, but their significance in dairying had not been clearly reduced to a 'help' position until roughly 1930.[21] The increasing masculinization of dairying, which the cheese industry initiated, had by the late nineteenth century encouraged male interest in more specialized 'commercial' dairy cows, but widespread attempts to breed cattle for higher milk yields, particularly in Canada, did not really gain much momentum until the 1920s. (It should be pointed out that although the Canadian ROP system existed for milk production in dairy cattle such as Holsteins, Jerseys, and even Shorthorns by the early twentieth century, specialized dairy cattle continued to be far outnumbered by specialized beef cattle until nearly 1920. Generally speaking, 'milch' cows, or milking cows, continued to provide limited amounts of milk. The slight rise in average milk yields, evident by the 1890s in the general Canadian herds, resulted from better feeding, not breeding for specialization.)[22]

We cannot know if men in general would have demanded specialized stock if they had been the main egg and meat producers early in the twentieth century in the United States and Canada. There is some evidence to suggest, however, that this would have been the case. The most prominent men in the American commercial poultry industry early in the twentieth century focused on the lucrative egg trade and undertook production on a large scale – Prentice, Johnson, Tancred, Hanson, and Kimber being classic examples. All five men, also serious breeders, used the single-purpose Leghorn. Male producers who were not breeders also tended to favour the egg industry and the single-purpose Leghorn for it. For example, a large Canadian operation called the Maple Leaf Poultry Farm, run by brothers in 1912 near London, Ontario, and described as a 'poultry ranch,' imported Leghorn chicks from the United States with the intent of keeping thousands of pullets for egg production, and selling the better cockerels for breeding purposes. (While a meat market would exist for the cockerel culls, this was clearly of secondary – or perhaps one should say tertiary – interest to these men.) The *Farmer's Advocate* stated that this was 'only the beginning' for this operation. 'Greater things are to come. The purpose is to go extensively into egg production.'[23] Men used Leghorns and ran large operations for egg production. Women tended to use dual- purpose birds and ran smaller egg-production operations.

There is other evidence, besides the fact that the sheer numbers of dual-purpose-type birds in the general poultry population reflected feminine domination of production, that women avoided single-purpose birds. In contrast to the Maple Leaf Poultry Farm's large-scale egg production with Leghorns, Canadian women reported other poultry breeding patterns to the *Farmer's Advocate* in 1912. When discussing their egg-laying operations and trapnesting experiences, they emphasized the use primarily of hens with a Plymouth Rock background, but also dual-purpose Rhode Island Red and Wyandotte lines.[24] W.R. Graham at the Ontario Agricultural College spent twenty years developing better strains of Barred Plymouth Rocks for farm operations. In 1916 he was asked, 'Why was the Barred Rock selected as the breed with which to work up a better strain of heavy layers for the Ontario farm?' Graham answered, 'Simply because there were more enquiries for cockerels of this breed and for eggs from this breed for hatching purposes than for all others combined. The farmer wanted Barred Rocks, so it was a good policy to work with what the farmer wanted. Correspondence has always indicated that the people of rural Ontario favored the Barred Rock as a farmer's breed.'[25] In spite of Graham's masculinizing of poultry production, the producer was almost always a woman before and after 1916 on Ontario farms, and she favoured the Barred Plymouth Rock, a dual-purpose breed, for egg production.

Feminine concern with dual-purpose stock or the use of birds in a dual-purpose manner, and the failure of poultry to be clearly divided by specialization until a late date did not mean that female poultry producer/growers took no interest in better 'genetics,' as so many farm-press editors implied before 1920 in both Canada and the United States.[26] Evidence may be spotty, but there is enough of it to suggest that many women were concerned with the quality of poultry they used – especially after 1900. Farm women clearly approved of what at the time was accepted as 'better' breeding, and bought from professional breeders as well, as female testimonials in the farm and poultry press of North America overwhelmingly revealed, and in spite of accusations by poultry breeders to the contrary. They might use cross-bred, or what should probably be described as grade, hens, but they often started with the best birds they could find. A notable number of women preferred to work with pure standardbred birds. One North Carolina woman campaigned for such birds and finally convinced her husband that it made sense to keep them. Some even borrowed money in order to purchase superior stock, and women normally paid close attention to what agricultural experts told them about the breeding of birds.[27] In spite of obvious female concern with good producing stock, poultry were often viewed as the poorest type of livestock on the farm. Decisive division by specialization held an unclear relationship with the idea of 'improved breeding.' Contemporaries linked female control to inferiority, and by extension were inclined to view lack of a specialization emphasis in poultry breeding as a characteristic, or even the outcome, of that linkage. Both women and men were blamed for this situation: women because of the type of bird they preferred and men because they took no interest in the breeding of their wives' chickens.

Women seemed to use the pure breeds of poultry in their production operations as much as men used purebred cattle and horses before 1920. Existing data on the actual numbers of fowl of pure breeds in Canada, for example, does not support the poultry breeders' contention that farm poultry were generally more 'scrub' or 'mongrel' in their background than any other type of farm animal at that time. The numbers of birds listed as pure match the numbers of horses and cattle listed as purebred when seen as a ratio to the total populations. The data suggests that the marketability of pure fowl matched that of purebred cattle and horses.[28] It might be noted that breeders of purebred cattle and horses in the United States and Canada were just as incensed over the low level of 'improved' animals of these species in the general herds as were poultry breeders over the situation of fowl.[29] No 'improved' breeding was as widespread as breeders thought it should be in any agricultural industry.

It is not known whether or not some male breeders of poultry felt hamstrung over the issue of purpose breeding in fowl because women – significant clients of breeders – wanted dual-purpose birds, thereby thwarting any interest breeders might have taken in a specialization emphasis. It would seem from Graham's experience, however, that poultry breeders were often prepared to bow to the prevailing demand, even if theoretically specialization interested them. Breeders discussed breeding egg birds separately from meat birds. The *Canadian Poultry Review* contained articles on the subject, written in 1892 by an American utility breeder. 'Eggs or Chickens – Which?' was the heading of one article. It depends on what is wanted, the writer stated, but answered the question by arguing that the emphasis should be on eggs because of the greater potential for profitability from them. 'For most farmers, to whom poultry keeping is a side issue,' he explained, 'the production of eggs is probably more profitable than the rearing of chickens.'[30] 'What Is a General Purpose Fowl?' was the title of another article addressing the problem by the same individual in a different issue of the Canadian journal.

> The general purpose fowl is simply a compromise – it takes from the great laying breed all the prolificacy which can be combined with the table qualities that it receives from the best table fowl. It is inferior in laying qualities to the best layers and in table qualities to the best table fowls; but it is superior in laying to the best table fowls and in table qualities to the best layers. It is better than either in the combination of the two qualities but inferior to each in its specialty ... As a matter of fact I believe it to be true that there are more general purpose fowls bred than there are of all other classes united. It is perfectly safe to make this assertion – the general purpose fowl is the most popular fowl in America.[31]

The problem, this person added, was that poultry keepers (and one would assume he largely meant women) argued that the general-purpose fowl was as good at one purpose as a bird bred to specialize in that use.

The Initiation of Specialized Breeding for Meat Purposes

The beginnings of a sizable North American demand for poultry meat in the 1920s initiated a new focus at least on the idea of specialization for meat purposes. In spite of the continued prevalence of a dual-purpose use of birds, the *American Poultry Journal* insisted in 1922 that the day of the true meat bird had dawned.[32] Exhibition breeders hoped to capitalize on this interest in meat-specialized birds. 'Since 1914,' the journal stated, 'the breeding and development of meat fowls has become almost a lost art, but at present it is again coming into a popular favour which bids fair to rival the popularity of high egg production.' The journal explained the way the fancy had worked with (or against) meat specialization over the years, and why abandonment of the exhibition-breeding system for meat birds was desirable. The Cornish was singled out as an example of the degeneration of meat-type birds. 'The Dark Cornish fowl is the extreme in meat type, the Leghorn the extreme in egg type ... The breeding of Dark Cornish, through the fancier, has degenerated into a faddism where fancy points have subordinated market worth to the background. Many specimens today appear more like feathered turtles than fowl ... Standard writers, judges and fanciers who have controlled the Standards have these many years been blind to the economic values of utility factors.'[33]

The day of the meat-bred bird had not in fact dawned, in spite of rising interest in meat production. In Canada, for example, while the number of Leghorns versus Rocks was rising, there was not a large enough market for meat to warrant a large-scale emphasis on the feeding of egg-oriented birds for that purpose, let alone the breeding of meat birds.[34] Demands for chicken meat were greater in the United States, and a number of American producers responded to that situation by feeding the egg-producing Leghorn for slaughter. Mrs Wilmer Steele of Delaware, credited with being the initiator of the modern broiler industry, decided to fatten her Leghorn pullet chicks in 1923 and sell them as meat rather than raise them as egg-layers. Others rapidly followed suit by feeding both pullets and cockerels. In 1925 Delaware produced fifty thousand meat birds and a year later sent over a million birds to market. Areas bounding Delaware in Maryland and Virginia began to raise birds for meat purposes too and the territory became known as Delmarva. By 1935 Delmarva was fattening seven million birds for market.[35] Many of these were still not from specialized efforts to breed for meat. That situation, however, was already undergoing change.

Some American craft breeders began in the 1930s to return to traditional ideas of cross-breeding for better meat production, and tried to create synthetic lines from cross-breeding programs.[36] In 1932 the Hall brothers of Connecticut introduced Rhode Island Red–Plymouth Rock cross-breds to hatcherymen and producer/growers in Delmarva.[37] The *American Poultry Journal* discussed the work of the Halls, noting the greater hatchability of cross-bred eggs (valuable to hatcherymen) and liveability of the chicks (valuable to producer/growers). The cross-breds also grew faster than standardbreds. One producer/grower who used Hall stock to raise broilers found, for example, that the cross-breds outgrew the Rhode Island Reds by 10 per cent and the Plymouth Rocks by 18 per cent at age four weeks.[38] Cross-bred stock, of course, implied terminal stock when it came to breeding purposes. Broiler growers of Hall cross-breds would be forced to return for replacements to suppler hatcheries who in turn had bought from multipliers of Hall-bred birds. Cross-breeding for hybrid vigour brought with it a biological lock. The cross-breeding experiments, however, led not just to the production of cross-breds and a biological lock. The Rhode Island Red/Plymouth Rock mix would result in the last new breed to be developed, namely, the New Hampshire, which was recognized in 1935 by the American Poultry Association.

S. C. Rhode Island Red cockerel, Champion, owned by the DeGraff Poultry Farm, Amsterdam, N. Y. This valuable specimen of this popular breed has won four first prizes and four color specials at Auburn, N. Y., Boston, Mass., Philadelphia, Pa. and Schnectady, N. Y., besides being the sire of many young cockerels promising to be his equal.

16 A prize-winning Rhode Island Red Cockerel in 1907. The Rhode Island Red was an American breed developed for utility and show purposes. It would become important for the modern broiler industry. Its shape is not unlike that of the Rocks, showing that there was a consistency of shape that went with good meat production, regardless of breed. (*American Poultry Journal*, July 1907, 627)

The Chicken-of-Tomorrow Contests and Specialized Meat Breeding

The separation of meat from egg purposes and the renewed focus on hybrids for meat set the stage for the most important event in the twentieth century for the poultry-meat-breeding industry, namely, the Chicken-of-Tomorrow contest, run by A&P (the Atlantic and Pacific Tea Company) Food Stores between 1948 and 1951. In 1945 Howard C. Pierce of A&P Food Stores addressed a meeting of Canadian poultrymen, saying that it would be a boon to the poultry-breeding industry if a uniform good meat bird could be developed. Considerable interest was taken by poultry concerns in this idea. A&P agreed to sponsor and underwrite a long-range project designed to improve meat-type birds. An independent committee (involving various poultry organizations, the poultry press, and the American Department of Agriculture) was formed to run a national contest in the United States. This group named itself the 'Chicken-of-Tomorrow' Committee and set up a structure for dressed-chicken-meat, state-based contests. A&P agreed to provide $10,000 in prizes, and the committee organized subcommittees to create score cards to assess the quality of entries in the contests, and wax models to illustrate what was wanted in a chicken carcass. A breeding subcommittee made recommendations to breeders.

In 1946, 44 Chicken-of-Tomorrow state committees held 31 poultry dressed-meat competitions. In 1947, the state committees ran 38 contests and five regional shows as well (with entries of the top five from each state within each of the five regions). A national competition in Delaware was planned for 1948; and 40 contestants, selected on the basis of 1946 state and 1947 regional results, entered. Each contestant provided two cases of eggs to a station in Maryland, where a hatchery hatched them. The chicks, moved to the Delaware Experiment Station, were placed in 40 separate pens. Reared there for 12 weeks and cared for under identical conditions, the birds were subsequently sent (as was normal practice in the chicken-meat industry) to a feeding battery for three days, and then slaughtered. Some 15,000 bird carcasses had to be judged. Fifty carcasses per entry, chosen on a random basis, were assessed. The score cards kept for each entry also addressed the issue of economy of production or feed conversion, egg production of parent stock, hatchability and liveability of the eggs sent to the hatchery in Maryland, average weight, rate of feathering, and the uniformity of stock within each pen. A second three-year set of contests were run between 1948 and 1951.[39]

In both the 1948 and 1951 contests, Charles Vantress won with a cross-bred Cornish/New Hampshire. Second prize in 1948 went to Henry Saglio of Arbor Acres with a pure White Plymouth Rock.[40] Saglio's Rock also won first prize for purebred fowl. Vantress concentrated on perfecting his cross-bred male lines and Saglio stressed his cross-strain female line. The breeds and breed-crosses that won the contest would be of long-lasting importance to the broiler breeding industry. The Chicken-of-Tomorrow contests established what might be described as the golden cross for meat production, namely, a Cornish/Rock/Red cross (or a Cornish/New Hampshire cross, the New Hampshire consisting of a Rock/Red cross) male and a White Rock–based female line. The contests did not teach how to breed, but rather showed breeders what to breed for, as well as confirming the fact that a viable and growing market for the right type of meat bird existed. Furthermore, the dressed-poultry contests gave the breeders a chance to compare their work to that of others, a situation that stirred up a sense of competition for improvement among them. Several breeders who were fundamentally unknown outside their local communities were skyrocketed into national fame, most particularly the two winners of the national contests.[41] The contests paved the way to complete acceptance of cross-breeding for hybrid vigour, and not the breeding of lines that reproduced truly, in broiler production. Such an approach to breeding ensured the presence of biological locks, and fit neatly with the entrenched geneticist attitude to breeding for agricultural improvement. The Chicken-of-Tomorrow contest, then, encouraged a geneticist approach to breeding, which in turn was attractive to corporate enterprise. Private breeders who had been successful in the contests found themselves at the head of companies who functioned increasingly with the aid of geneticists.

The Development of Integration in the Structure of the Meat Industry

At the same time that experiments with cross-breeding for meat birds were under way, accompanied by the rise of corporate breeding, the structure of the meat industry itself was undergoing extensive change. The connection of the feed industry to the broiler industry initiated what would ultimately be almost complete vertical integration of the meat industry outside the activity of breeding. Traditionally, feed companies in the United States and Canada had operated as grist mills, grinding grain which farmers brought in and marketing by-products from this work as feed. Farmers tended to mix the product with feed grown on the farm. Greater attention to nutrition and the effects of vitamins, as well as evidence that their use in feed enhanced growth and good health, changed the situation, and encouraged feed companies to market their products more aggressively. By the late 1920s, farmers in increasing numbers stopped relying on feed grown on the farm.

The rising fortunes of the feed companies led to American branch planting in Canada. In the 1920s a number of large American-based feed companies moved into Canada to take advantage of the growing potential market. For example, Ralston-Purina, after selling feed for some years in Ontario, bought a cereal mill in Woodstock in 1929. The Canadian Feed Manufacturers Association was formed that year by Master Feeds, Quaker Oats, Maple Leaf Mills, and Ralston-Purina, thereby better organizing the interests of the feed industry in Canada. Changes in the Canadian Bank Act of 1954 made it easier for the American feed companies to provide credit to poultry farmers, thereby financing much of the Canadian poultry-meat industry. By the mid-1950s the feed companies contracted with poultry keepers growing meat birds (who were increasingly men), and shared both the cost and profits when marketing time came. If the farmer could not carry his share of the loss, the feed company would be forced to do so.[42] One way to assess feed company power in the Canadian poultry industry generally is to note the size of their ads from the 1930s to the 1950s in the farm press. Always large and often covering four pages by the 1950s, Ralston-Purina's ads offered anything a poultryman would need.[43]

As early as the 1920s, though, processors had started to play a role in the integration movement of the chicken-meat industry, a pattern that would be more pronounced by the 1970s when the feed companies began to divest their broiler business interests. In Canada, for example, the packing company Canada Packers marketed meat by-products in the 1920s as feed under the name of Gunner. Swift Canadian (a subsidiary of the large American packing company Swift & Co.), already the owner of a hatchery in Ontario, went into the feed business and set up poultry processing plants nearby their hatchery.[44] The trend to processor control and away from feed-company control in both the United States and Canada would be encouraged by developing government inspection regulations relating to grading and health issues.[45] North American fluctuating broiler prices in the 1950s also played a role in the move to the processor as integrator. Feed companies found the broiler industry economically unattractive under these volatile conditions, and so began to get out of the broiler business in both countries.

Ralston-Purina, for example, sold its broiler interests to the huge American processor Tyson Foods in 1972.[46] The processor's better ability to control the retail marketing of the product to the consumer seemed to make such an investment worthwhile.

Complicated issues relating to the crop lien system in the American south encouraged the development of certain integrative patterns in the broiler industry that evolved in states like Georgia. In the 1930s early integrators, normally merchants, began to extend credit to upcountry Georgia cotton farmers, hard hit by the devastation brought about by the boll weevil, who turned to the raising of chickens. (Between 1921 and 1923 the Georgia cotton yield per acre had dropped more than 50 per cent.) The move to more extensive poultry farming came naturally to these families, because the women in particular understood chicken husbandry, having been involved in poultry production for generations. Stimulated by the Depression, southern farmers increasingly abandoned cotton planting on their land in the 1930s, joined their wives in poultry farming, and became dependent on an expanded chicken enterprise which supported a new one-crop economy. Emphasis shifted to complete commercial production, as integrators discouraged the keeping of a separate farm flock designed to sustain the family. Georgia chicken producers still maintained some freedom with respect to markets throughout the 1930s, even though financed by merchants and feed companies. They could sell the finished bird to processors of their choice.[47] Integrator control over farmers raising chickens advanced considerably in the 1940s, at the same time that cross-bred meat-producing chicks became increasingly common. Chicken farmers contracted with feed companies and sold to processors stipulated by the feed company. Farmers still paid the feed companies for the chicks (and therefore owned the stock) and feed, doing so after the sale of the birds. Feed companies extended credit to the farmer in the form of chicks and/or feed.

Georgian broiler-integrating feed companies did not have to be large when they entered the poultry industry. The story of J.D. Jewell of Georgia was a case in point. In 1936 he decided to provide baby chicks to local farmers, along with feed from his failing feed business, and subsequently to pick up the finished broilers from farmers and sell the birds himself. Farmers would receive some of the selling price in return for growing out the stock. A local bank financed the buying of the chicks from a hatchery. By 1940 Jewell had bought a hatchery to produce his own baby chicks and began to construct a processing plant.

He now owned the baby chicks, contracted farmers, processed the broilers, and then marketed them.[48] Jewell's success echoed a continuing rise in broiler growing in poverty-stricken areas of the South that occurred after the Second World War. Between 1948 and 1953 Georgia increased production by 143 per cent. Between 1947 and 1960 the area of the south-east grew in production rates by 365 per cent.[49] Changing patterns in broiler integration did not bring breeding into the structure.

The Second World War played a role in encouraging the growth of the meat industry over the egg industry in the United States. The war stimulated the American broiler industry directly, because of internal domestic factors. Wartime measures encouraged the greater domestic consumption of chicken, and these ultimately triggered further a poultry industry centred in the South. Unlike beef, chicken was not rationed. The government encouraged the consumption of chicken meat in order to keep the more desirable meats like beef available for the troops. Raising and eating chickens became a patriotic duty.[50] When this pressure to eat chicken combined with the way the government restricted sales of the major poultry-meat centre, the situation was ripe for change. In 1942 Delmarva produced 90 million meat birds a year – more than any other region (the future centre, Georgia, sent 10 million birds to market that year). In 1942 the War Food Administration declared that all Delmarva production should be made available for the armed forces. While the American government's action guaranteed Delmarva solid markets for the duration of the war, in the end the move aided Georgia and other areas of the South. These new producing areas managed after the war to overcome the advantage that Delmarva had always had, namely, greater proximity to the major urban markets. When the army cancelled its contracts to buy Delmarva products in 1946, producers in the South had captured the home markets, at the same time that more Americans had learned to eat chicken.

In Canada the war initially stimulated egg production. The 'Eggs for Britain' campaign, 1941–6, greatly enhanced the Canadian egg industry and encouraged its growth. With the loss of the British market in 1949, when Britain ended the special agreement to buy Canadian poultry products, potential overproduction forced poultry producers to concentrate on a home market.[51] The number of chickens in Canada had grown from 56 million in 1939 to 74 million at the end of the war. By 1950 the numbers had dropped to 56 million.[52] The Canadian chicken situation changed rapidly over the 1950s, as the broiler industry took centre position. By this time, much of Canadian industry independence in both the egg and broiler industries had been undermined. The trend had begun with the influx of American egg-laying genetics to franchised hatcheries, but grew rapidly over the 1950s, particularly in business organizations relating to the meat industry.

The franchising of egg-laying chicks to hatcheries was matched by the purchasing of hatcheries by feed companies who wanted to streamline their broiler interests. By the late 1960s over half of Ontario's hatchery capacity, for example, was owned by feed companies.[53] Chicken producers of meat birds, now generally men, also lost independence at an increasing rate in Canada just as they had in the United States. In 1959 it was estimated that 66 per cent of broilers in Ontario were being raised by farmers under such contracts to feed companies.[54] The feed companies Ralston-Purina and Pillsbury, in particular, dominated the shifting integrative involvement of the feed industry with the American broiler industry. Companies such as Ralston-Purina and Pillsbury controlled 90 per cent of American broiler production by the 1960s. They owned the hatcheries which produced baby chicks, feed mills, processing plants, and by this point the birds themselves as well. Farmers were contracted to grow out the stock owned by the feed companies.[55] Farmers were responsible for the most expensive part of the business – all equipment needed to house the birds, and all labour.[56]

It has been argued that the North American poultry industry of the 1950s was largely built by the poultry-meat industry.[57] Technological developments had advanced sufficiently by that time to enhance the industry's capacity to undergo ever-increasing expansion. A greater understanding of the important role that both protein and vitamins played in the health and growth of birds existed by the 1940s. From the 1940s to the 1960s new knowledge of poultry nutritional requirements allowed for faster production of poultry under any conditions.[58] Better vaccines evolved. The production of superior birds in large numbers which could utilize these advancements with greater efficiency became possible under these conditions. Domestic consumption rates of poultry products rose accordingly in both Canada and the United States, and they did so most dramatically for poultry meat. More than three times as much poultry meat was consumed in Canada in 1960 than in 1921, while egg consumption had risen only slightly.[59] These are not per capita figures. They reflect total domestic consumption, but consumption of rising population numbers. Even so, the data makes it clear that the rise in egg consumption was no match for the rise in poultry-meat consumption in Canada. In the United States, the per capita rate of egg consumption actually fell slightly between 1940 and 1965. In comparison, per capita consumption of poultry meat rose 16 times between 1940 and 1965, from 2 pounds to just over 30 pounds.[60] Chicken had become available to a greater degree by 1960 than probably even Herbert Hoover dreamed when he made his famous campaign slogan in 1928 – 'A chicken in every pot.'

As the production and marketing (but not breeding) activities within the broiler industry across North America became integrated, contemporaries saw clearly that these patterns presented the manifestations of a revolution. The changes that integration brought to the industry exhilarated, concerned, and worried poultrymen. At various Ontario Poultry Conferences held at Guelph, geneticists and poultrymen from both the United States and Canada discussed the North American integrative process generally at considerable length. In 1957 the subject dominated talks. C.K. Laurent from Georgia spoke of both egg and broiler integrative patterns in North America and argued that integration promoted efficiency. Integration could occur in many other ways; between truckers delivering baby chicks and hatcheries who produced them, for example.[61] An article in the *Farmer's Advocate* in 1959 did not see integration in quite such a glowing light.

> Trends in the Poultry Industry – Integration, an unknown term in agriculture a couple of years ago, has entered several branches of agriculture, and the poultry industry with a vengeance ... The present trend depends on volume. The farm flock of a few years ago has been bypassed by mass production. Three firms were reported to be now processing 60% of the broilers produced in Canada ... [But] it is a tough deal for a lot of producers and a case of survival of the fittest. The number of hatcheries was reported to have decreased by one half and feed firms and processors are consolidating.[62]

Integration continued to provoke conflicting sentiments. In 1964 a major Canadian breeder, D. McQueen Shaver, stated at a general poultry industry meeting that integration encouraged profit to the overall egg and broiler industries by lessening conflict between various segments. J.M. Appleton, vice-president of a large feed company, agreed. He described integration as cooperation of the producing, processing, and marketing arms of either the egg or broiler industry, and argued that integrationists believed that all sections of an industry must prosper together if there was to be any lasting success. Appleton meant all sections except the breeding section, which remained separate from the rest. But what about the welfare of the poultry producer and what about the future of the family farm, he asked? 'We have come face to face with a juggernaut of our own making and we must bring it under control before it destroys [the industry] we have been building for the last 100 years.'[63] Standardization, mechanization, and subsequently industrialization through business channels, this individual recognized, brought with them fundamental changes to farming practices that would profoundly alter the very nature of, not just food production, but rural life as well.

Integration and Changes on Farms Producing Chicken Meat

One important shift that accompanied integration was the move to larger and larger producer/grower operations. I.K. Felch had estimated that in the 1870s American flock size varied from 12 to 50. Larger ones were extremely rare.[64] The flocks on some American farms had risen by 1913 to between 100 and over 200 hens.[65] (They tended to be larger in the Midwest than the South.)[66] Even so, the vast majority of commercial flocks until after the 1950s continued to number below 200. As late as 1930 the average flock in the United States remained about the same as Felch's estimate for the late nineteenth century. Over half of the farms reporting chicken keeping in 1930 stated they had fewer than 50 hens, and the average number of hens per farm was estimated to be no higher than 23.[67] Flocks in Canada over this period were just as small. An informal survey for Ontario done by *Farming World* in 1901 set the average number of birds per farm at 79.[68] A national census undertaken in 1903 by the journal gave an average of 56 per farm, but many believed that 30 was a more representative number for a country average.[69] The average number of laying hens per farm had not risen appreciably in Canada by 1924: set at 68.[70] Flocks that substantial were anomalies. A general trend to larger flocks had not yet begun by 1950.

Farm experts seemed to sense both the coming of an increased number of significantly larger-sized flocks and the important impact they would have on the poultry economy before either trend had started. As early as 1930 William Graham, poultry expert at the Ontario Agricultural College, concluded that growth in flock size, while not apparent yet, would happen and change the dynamics of chicken farming. 'I sometimes wonder whether, in the end, we are going to raise chickens on the farms or in chicken factories. I am suspicious that we are going to have chicken factories,' he said.[71] Nothing really changed, though, for some time in Canada. In 1939, for example, the *Farmer's Advocate* pointed out that flocks of well under 200 birds dominated 75 per cent of Canada's chicken population. 'It is claimed by men in a position to know that the number of flocks of 500 birds and over or those generally classed as commercial flocks constitute only 5 per cent of the total flocks in Ontario and do not exceed 3 per cent in all of Canada,' the journal added.[72] By 1955, poultry experts at the Ontario Agricultural College saw that Graham's prediction was going to come true: the days of farm flocks averaging fewer than 200 birds were numbered.[73]

It was evident by the 1960s that such flocks were less common in the poultry producing-world, and more importantly, they played a reduced role in the chicken economy. It was estimated that 62 per cent of poultry products on the American market in 1964 came from farms with flocks that could number in the tens of thousands, or as high as 100,000 when it came to a concentration on broiler production.[74] Various factors allowed for this huge increase in flock size: namely, changing technology in housing, feed, and health.

The Broiler Industry and the Changing Position of Women in Chicken Farming

The rise of a significant number of large-sized farm flocks, their increasing dominance of the commercial industry, and an emphasis on meat production coincided with the rise of male labour in, and ultimately control of, the most important commercial aspects of chicken farming. (The existence of larger commercial poultry plants, evident in isolated cases as early as about 1910, coincided with male interest, as the farm press made clear.) Women might still work with chickens after 1960, but they did not run these large poultry operations.[75] When meat production became thoroughly separate from egg production and hybrids replaced purebred birds, men replaced women as the major producer/ growers in commercial poultry farming in both the United States and Canada. Small and often female-run flocks might continue to exist, but they became increasingly less common after the 1960s and played a continually less important role in the economy of both the meat and egg industries as well. Integration encouraged all these trends.

Changes in the North American broiler industry occurred simultaneously with the gradual removal of women as producer/growers from the poultry industry generally and the poultry-meat industry in particular. What the connection might be between the phenomena – namely, the replacement of women by men as producers of meat birds and the shift to clear division by specialization in breeding – is not clear. The increased commercialization that went with the new emphasis on meat production, though, must have played a central role in these changes. A move to mono-culture with respect to poultry production seems to have been another aspect of the shift.

Corporate Breeding and Its Culture

The masculinizing of production enhanced the move to making chicken-meat breeding a 'scientific' endeavour, and the subsequent changes in broiler breeding – namely, cross-breeding for hybrid vigour and biological locks – confirmed that the complete demise of a North American breeding culture, entrenched since the mid-nineteenth century when the commercial egg industry was born, had taken place. The emphasis on perpetuating lines that bred truly had ended. The rapid development of the broiler industry, underpinned by specialized meat-yielding birds of a cross-bred background, matched the revolutionary changes in breeding that had occurred in the egg-breeding industry. This major shift in outlook, present now in all chicken breeding, brought with it to North American breeding the introduction of corporate involvement, and encouraged the entrenched separation of the breeding industry from the rest of the integrated poultry industry. Breeding for heterosis allowed breeder organizations to protect their products from other breeding companies as well as integrator companies by keeping the breeding lines used to create them a secret. The hybrids that served the broiler industry and that resulted from crossing for heterosis were useless as breeding stock. One had to access the parents or, more likely, the grandparents to recreate producing hybrids. By the 1950s the breeder companies no longer released even one sex of such stock for breeding purposes. Breeders ceased marketing any genetics for breeding purposes. They sold only non-breeding material. The trade in genetics in both countries, therefore, had ended. Breeders named the strains that constituted the parent and grandparent stock by number or secret code. No published records exist as to how selection proceeded to create the lines at that time. The 1950s saw the introduction of calculating machines (or what might be called rudimentary computers or at least data processors) which allowed for the far better use of quantification.[76] Breeding for hybrid vigour, the use of calculating machines, combing the countryside for potentially hidden strains that might still exist and be used for crossing purposes, and increased secrecy were all features of mid-twentieth-century poultry breeding for both eggs and meat in North America. The situation introduced an ongoing pattern: namely, the virtual end of trade in genetics, a reduction in actual breeders, and the tendency of the breeding industry to consolidate into large companies.

By the late 1950s poultry geneticist breeders involved in both egg-laying and meat-producing breeding often lay in two camps at the same time – that of scientist and practical breeder. The situation could exacerbate the underlying tension that had existed from the time of Mendelism's appearance in 1900 between scientists and traditional breeders over the matter of livestock breeding (and, one might argue, the problem of hegemony over hereditary knowledge that had existed from the time of Darwin, and was clearly articulated by W. Johannsen as early as 1911). I.M. Lerner serves as an example of the phenomenon of heightened conflict over the input of scientist and practical breeder to the problem of agricultural breeding when one person was both a geneticist and a breeder. All his life he utilized chickens as laboratory animals in order to understand the process of genetics, or what he might have called the genetics of the fowl, and what that information had to offer to the study of evolution.[77] But Lerner used his breeding experiments to study the effectiveness of various selection methods on the productivity of hens, and in doing so aided the developing quantitative and scientific poultry breeding that was evolving rapidly by the late 1930s. In the 1950s he wrote several books designed to transform all animal breeding 'from an art to a science based on multifactorial Mendelian inheritance,' as R.W. Allard claimed after Lerner's death.[78] In 1966, along with H. Donald, who was co-director in the 1950s of the Animal Breeding and Animal Genetics branches of the institute devoted to agricultural breeding at the University of Edinburgh in Scotland, Lerner wrote a third book, *Modern Developments in Animal Breeding*. This work addressed the issue of science (which the authors saw as empirical thinking) and art (which for the authors meant emotional thinking) in the breeding of farm stock. Lerner and Donald stated: 'The business of breeding [purebred] stock for sale is not just a matter of heredity, perhaps not even predominantly so. The devoted grooming, feeding and fitting, the propaganda about pedigrees and wins at fairs and shows, the dramatics of the auction ring, the trivialities of breed characteristics, and the good company of fellow breeders, constitute a vocation, not a genetic exercise.' At the same time that purebred breeders continued to reject quantification (and still know nothing, as a result, about how to pick good productive stock), the authors stated that

> poultry breeding began its astonishing metamorphosis at the hands of a combination of geneticists and businessmen ... In [the poultry industry] geneticists-mangers have taken over the direction of breeding policy in most if not all advanced countries ... But, in the larger classes of livestock, the biologically untrained men are still in charge and the differences in background, outlook, and in sophistication between them and the academic or research biologists, needless to say, leads [sic] to mutual antagonisms ... It is possible that geneticists underestimate many practical difficulties in their simplified theoretical concepts, but it is certain that many breeders refuse to understand and to credit geneticists having advanced their cause in any way ... The usual question asked by the breeder of the scientist is what genetics has contributed to animal breeding.

The answer, the authors said rather unclearly, was that one must think scientifically (or empirically), not emotionally (or artistically).[79] Lerner and Donald were inclined to view the breeding of lines that perpetuate themselves truly, as purebred breeds do, as antagonist to the scientific way to breed, which relied on terminal crosses for heterosis. Breeding for lines that bred truly seemed in a hazy way to the authors to mean breeding artistically.

The Dutch geneticist A.L. Hagedoorn serves as another example of a scientist dismissing the work of contemporary practical breeders. Increasingly by 1950, he viewed cattle breeders in the Netherlands as being backward in their breeding approaches. Many Dutch agricultural experts believed he had little appreciation of the conditions under which cattle were raised on farms in Holland. But while he might not have always understood the difficulties that breeders of these large livestock faced in the mid-twentieth century, he was well aware that 'genetic' breeding of livestock had firm foundations in craft breeding knowledge. Trained by de Vries and a Mendelist, Hagedoorn argued that geneticists until at least 1940 owed virtually all of their knowledge on large livestock breeding to the accumulated knowledge of the craft breeders.[80] Hagedoorn worked closely with chicken breeders who were exhibition or commercial breeders in both Holland and Britain until his death in the 1950s.[81]

In 1960 A. Fraser, a British livestock expert and senior lecturer in animal and dairy husbandry at the University of Aberdeen in Scotland, published *Animal Breeding Heresies*, in which he attacked this righteous attitude of geneticists towards untrained breeders. 'The breeds at our service,' he wrote, 'the basic stuffs at our command; the husbandry systems in which these are profitably joined, were all well established and confirmed before the first agricultural scientist had been appointed ... The scientist might well be advised to go occasionally to the farmyard to learn rather than teach, or, what is far less excusable, to preach.' Fraser spoke as well of the way genetics interfaced with livestock raising:

> In addition to its excessively mechanistic approach to living things, modern science, at least as applied to agriculture, threatens to become dogma. That seems to me to be particularly true of the science called genetics. The geneticist appears to be perfectly prepared, even eager, to impose his theories which he would appear to believe infallible, upon the livestock industry of consenting countries by continuous exhortation, if necessary reinforced with legislative action. His theory, so closely in danger of becoming a creed ... is the accepted science in Western countries.

Fraser summed up his sense of geneticists in the following words: 'I feel, therefore, that the traditional knowledge based on practical experience which forms the solid basis of our animal husbandry, through the criticism of certain scientists a trifle too wise in their own conceit, is being ignored or even somewhat unfairly derided, and that the newer knowledge based on science is being applied too uncritically and with too great haste.'[82]

Chapter Seven

Epilogue: Trends in Chicken Breeding after 1950

The fundamental characteristics of the chicken-breeding industry that had been laid down by the 1950s remained in place over the last half of the twentieth century. The breeding industry demonstrated patterns which might be described as the continuation, consolidation, or logical extension of features well entrenched by the 1950s. The most significant shift that took place over that period in chicken breeding was the globalization of the undertaking. By the early twenty-first century, however, there are signs that we are at the beginning of a major revolution which would see the decline of hybrid breeding, a return to lines that breed truly, and the initiation of selection methods based more on DNA markers than on the progeny test. In the process, molecular genetics may largely replace quantitative or classical genetics. Increasingly powerful computers will play a role in that revolution.

This final chapter provides a general and brief overview of developments in the chicken-breeding world from the 1950s to the early twenty-first century. I first review the growth of the two chicken industries, and outline the consolidation of breeding companies within that world. I next discuss breeding trends and the breeding industry from the 1950s into the twenty-first century. Finally, I describe the work of one Canadian breeding company, the Shaver Poultry Breeding Farm, from roughly 1960 to 1985 in order to illustrate how secrecy worked, what contributions scientists made to breeding programs, how business acumen affected the welfare of the breeding companies, and the way the breeding arm fit into the rest of the poultry industry. I conclude by describing Canada's Management Supply System in relation to broiler production.

Globalization of the Broiler and Egg Industries

The phenomenal growth of the broiler industry within North America in the mid-twentieth century would have ramifications around the world. Glut and oversupply, evident by the late 1950s, and serious by the early 1960s, compelled the American industry to look for new markets abroad as an outlet for products that could not be sold at home. Industrialized agriculture dictated a greater focus on global markets. The export of American poultry products to Europe had begun in the early 1950s as part of trade agreements relating to assistance programs following the war, but the pattern accelerated with business troubles at home. For example, four million pounds of poultry meat was shipped to West Germany in 1956, and by 1963 that number had risen to 152 million pounds. Shipments from the United States to France, Italy, and Belgium-Luxembourg were also on the rise. By the 1960s, however, much of this trade contravened the GATT (General Agreement on Tariffs and Trade) regulations, set in place by the EEC (European Economic Community), forerunner of the EU (European Union). The EEC had been established in 1957 in Rome by West Germany, Italy, France, the Netherlands, Belgium, and Luxembourg; and economic conflicts within it matched those outside it and in its relations with other European countries. The United States was drawn into the conflict because of its massive exportations to the EEC. The result was an international and transatlantic trade conflict, known as the Chicken War, which took place between 1961 and 1964. The problem had subsided by 1967, with imports from the United States into the EEC down by 45 per cent from their 1961 level.[1] The resolution of the Chicken War did not result in lasting trade barriers for chicken products. Enormous changes in the business structure of the poultry industry and the poultry-breeding industry that occurred after 1970 would enhance the tendency to globalize the poultry industry.

The per-capita rate of poultry-meat consumption would also continue to rise worldwide. In the 1960s, chicken had become part of the fast-food chain with Kentucky Fried Chicken, but it was McDonald's Chicken McNugget, introduced in 1983, that expanded the demand for chicken meat. Processed chicken quickly became more important than chicken as a whole or as parts of that whole.[2] By 1995, for example, Americans consumed 50 pounds per capita a year, and by 2000 that figure had climbed to about 80 pounds.[3] McDonald's extended this American trend to greater chicken consumption by selling processed chicken as McNuggets in its 30,000 restaurants in over 100 countries. Eating processed chicken has become a global phenomenon.[4] At the same time that these developments have been taking place, there has been an increased demand for eggs – also to go into processed foods, and not simply for consumption of the egg itself.

The North American poultry industries for both eggs and meat have increasingly become ones of astonishing complexity. Evolving science and technology over the past fifty years have both continued to play a major role in that development. Other factors are part of the story. The growth of government agencies to inspect the quality of the egg or meat product, smoother marketing agents, effective advertising, faster transportation methods (primarily via aircraft), uniform packaging, and the rise of better slaughtering facilities for poultry meat had to be in place before the birds could supply the eggs and meat they do, in a cheap and safe manner and in enormous volumes. It is indeed a daunting thought to grasp, but each and every one of these factors had to increase in sophistication at the same pace in order to make what would become global egg and meat industries work the way they do today.

Within this complex framework of corporate structure and business growth that evolved in North America, poultry breeding remained separate from the process of integration. As integrators took over virtually all aspects of the poultry-meat industry and many aspects of the egg industry in both the United States and Canada – even to the production of baby chicks at hatcheries – they did not generate the breeding that went into those chicks. The story of poultry's rise to agribusiness (first in North America and then globally) is, therefore, not complete without an understanding of the transition that took place in a separate industry, namely, the breeding industry.[5] As early as the 1950s, North American breeder companies would find, like the integrators, that they needed larger markets to survive, and the result was branch planting in Europe. When the companies began to operate on a more global scale, poultry-breeding practices around the world became increasingly more homogeneous.

Consolidation of the Modern Breeding Industry

In the 1970s, although the breeder sector continued to remain separate from other industry sectors, the number of independent companies became fewer and more international when drug companies – Upjohn, Pfizer, and Merck Drug Company, for example – began to collect and consolidate various operations. The drug companies, often veterinary in orientation, hoped to capitalize on markets for vaccines, a strategy that had worked in the plant industry with the sales of chemicals for pest control.[6] In 1980 there were about twenty breeding companies worldwide serving the broiler and table-egg industries. Consolidation trends indicated a move away from North American control of the world's poultry-breeding industry and towards European control. As early as the late 1970s European drug/veterinary companies began to buy various North American breeding concerns. Institute de Sélection Animale (ISA), founded in 1976 in France, began to collect breeding companies, including the significant Canadian breeder company Shaver Poultry Breeding Farms in 1989. Merck Drugs came to own the Shaver operations through a company named Merial Animal Health, created by Merck and French concerns in 1997, which bought ISA.[7] The Erich Wesjohann Group, a privately owned company headquartered in Germany, acquired Hy-Line International and many others which operated either as wholly owned subsidiaries or as joint ventures. Some breeding companies were absorbed into the major breeding company Cobb-Vantress, which in turn was bought by the integrator Tyson Foods. This move linked the primary breeder to the integrator, even though Cobb-Vantress retained some independence as a subsidiary of Tyson Foods. Cobb-Vantress sells genetics to Tyson and to others as well. By 2007 more that 90 per cent of breeding stock used for broiler production came from three breeding companies. Consolidation of breeding companies has not resulted in a reduction of genetic diversity. More actual products are available from the few existing breeding companies than had been true when a greater number of broiler companies were producing.[8] This situation can be explained by the fact that contraction of breeding companies actually allowed for the revival of 'trading' in genetics. Large holding corporations of subsidiary breeding companies can utilize a wide range of genetic resources because they own them. Smaller independent companies, even if there were more of them, had access to a much more limited genetic base.

The paths that the ownership of Arbor Acres and Ross Breeders took reveal how mergers financed by venture-capital companies dominated the corporate shrinking of the breeding industry by the beginning of the twenty-first century. Ross Breeders Ltd, a primary broiler breeder that had originated in Scotland in the 1950s, and the primary American broiler breeder Arbor Acres went into a joint venture in the 1970s in which Ross was to produce a good male line from its Cornish hybrid strains, and Arbor Acres to produce its female White Rock lines, thereby utilizing the golden Rock/Cornish cross promoted by its success in the Chicken-of-Tomorrow Campaign. The venture failed and the two companies were bought separately by a variety of organizations (by the early 1990s Ross was owned by the British company Hillsdown Holdings, and Arbor Acres by the British company Booker Food). B.C. Partners, a London-based venture-capital concern established in 1986 and a company with an interest in Ross Breeders by the end of the 1990s, decided to invest more heavily in the international poultry-breeding industry.[9] B.C. Partners acquired Arbor Acres, and formed Aviagen in 2000 by combining the Ross holdings with the newly bought interests in Arbor Acres. Advent International, a global private equity investor, acquired Aviagen in 2003 from B.C. Partners. In 2005 Advent sold its interest in Aviagen to Aviagen's strategic partner and minority owner, the Erich Wesjohann Group. This operation had become a strategic partner in 2003 through the sale of its holdings in Ross Breeders for 21 per cent of Aviagen. Under its new ownership, Aviagen was to function independently. Its only duty to the group was to be profitable. By this time, both Ross and Arbor Acres had ceased to exist as separate entities. The names referred to lines produced by Aviagen from strains that had arisen primarily from the Ross male and the Arbor Acres female.[10]

Genetics and the Breeding Industry, 1950s to 2000

While breeder companies, under the direction of trained geneticists, all continued to focus on some form of cross-breeding for heterosis, the actual practices (not the theoretical methodology) followed by egg and broiler breeders, as well as the lines they used, remained hidden.[11] North American breeder companies continued to work with the idea that keeping breeding secrets enhanced business control over intellectual property embedded in the genetics of their lines against other breeding competitors.[12] Most information that we have on breeder-company approaches to breeding and selection arises out of scientific literature which, while it provided the results of certain breeding experiments, did not reveal what was actually done by breeding companies. The geneticist/breeders presumably used the information to modify breeding practices within their companies. Over the 1950s and 1960s, breeder/geneticists in North America met at conferences to discuss their problems, and it was intended that breeders share general strategic ideas. This did not happen to any great extent.

A serious competitor of the hybrid-corn breeding method for the breeding of egg-laying hens emerged the 1950s. The potential for a strategy that could overcome some of the problems inherent in the hybrid-corn method – its cumbersomeness and its cost – had attracted the attention of breeders since the 1940s. One particular individual came up with a methodology which overcame those problems. Arthur Heisdorf, who had been interested in cross-breeding and heterosis as early as the 1930s and had worked as a geneticist with Kimber Farms, established his own breeding operation in 1945 under the name of Heisdorf and Nelson. In 1950 he began to experiment with a system called Recurrent Reciprocal Selection (RRS) breeding, which relied on nicking alone.[13] Selection for breeding under this method did not rest on the development of synthetic or pure lines, but solely on the ongoing attempt to improve the good nicking ability, or combining ability, of two specific strains (or lines). The developing strains stayed in a constant state of change because they were judged at every generation only on how well they had crossed on each other.

Theoretically, this approach differed profoundly from that which governed the establishing of synthetic or pure lines. If better heterosis was desired from the crossing of synthetic or pure lines, then new true-breeding lines had to be created. With RRS, the existing strains continued to be used, but were altered by selective breeding in order to make them meet better nicking standards. RRS could be used within one breed, or from lines that combined a number of breeds. The method was less cumbersome and wasteful than the hybrid-corn breeding method, which carried many synthetic or pure lines at the same time and constantly created new ones, in order to test the best heterosis cross. It was also elastic as to ways of selecting stock to increase nicking capacity. Because it was less expensive, it could be undertaken by smaller egg-laying breeders, and it was the hope of some that RRS would allow them to continue to operate within an increasingly corporate breeding world.[14] Many geneticists in breeding companies, however, remained sceptical about the value of RRS until into the 1960s.[15] Not being lines that bred truly, RRS strains tended to be more heterozygous than synthetic or pure lines, meaning that, theoretically, they would produce variable and therefore unacceptably unpredictable results. Some geneticists, however, did believe that Reciprocal Recurrent Selection worked.[16]

By the early 1970s the hybrid-corn breeding method was on the way out in the production of layers, at a time when many systems involving intense inbreeding became unpopular. Breeding for layers continued to be aimed at finding good combining ability of lines, however, and as many as eighty might be kept for crossing. Experimental lines, as well as breeding lines, were maintained separately. Because the traits desired in the egg industry are only expressed in females, males continued to be selected on the basis of the performance of their relatives – their progeny, their half-siblings, or perhaps the progeny of their male relatives. Breeders of layers did not simply look for better egg production in females, but rather selected for multiple traits – egg production, fertility, hatchability, egg weight, shell colour, body weight, sexual maturity, and feed efficiency. While the progeny test played a large role in selection for layers, mass selection continued to be used.

The North American egg-breeding industry was hampered by a special problem, which compelled breeders to work within one breed, and to cross only strains within that breed. The overwhelming popularity of white eggs with consumers in North America enforced an emphasis on Leghorns and the avoidance of cross-breeding, because the Rocks and Rhode Island Reds produced brown eggs and crosses with Leghorns often resulted in eggs even less desirable – coloured somewhere in between and described as 'tinted' eggs. (New England continued to be the only major centre favouring brown eggs.) Leghorns responded well to selection for distinct lines within the one breed. It was easier to create separate strains of Leghorns, and maintain good vigour, than it was in the heavier breeds like the Rocks. The Leghorn, it was assumed, carried a smaller genetic load, meaning it had fewer potential lethal recessives that could surface in a homozygous state than did the Rocks.[17]

Breeders of broilers also emphasized the progeny test, but mass selection still played an important role in their programs. Like table-egg breeders, they selected for a variety of traits and kept a number of lines and, like table-egg breeders, were forced to face different selective factors in males and females. Rapid weight gain and good feed conversion tends to decrease fertility, which is a problem in the reproduction level of females for the broiler industry. Female lines were selected on the basis of growth and conformation (as were males), but attention was paid in females as well to egg production, fertility, and hatchability.[18]

By the late 1970s it was evident that practices used to achieve heterosis in both egg layers and broilers were undergoing change. Most systems involving intense inbreeding had become unpopular in both poultry-breeding sectors. Computers had made quantification and assessment of various lines' productivity in egg and meat breeding easier.[19]

The relative importance of a certain characteristic against another and the effectiveness of any given selection system within that framework could be assessed with considerable accuracy.[20] All breeders of both egg layers and meat birds, however, were more interested in how they compared with their competitors than in actual improvement of their breeding programs.[21]

The advancement in computer power since the late 1990s allowed the older methods of selection to work with quantitative genetics better than they had in the past for heterosis, by making the quantification of ever more complex sets of characteristics in birds bred under such strategies possible.[22] By the twenty-first century breeder companies producing broiler genetics were able to concentrate on selection for traits involving heart/lung fitness, feed conversion, growth profile, immune response, meat quality, breast meat, weight, hatchability, eggs, and skeletal integrity, and do so without resorting to inbreeding. Each of these sectors included several distinct traits. Breeding for some traits could be antagonistic to breeding for others, and modern geneticist poultry breeders understand that they must strategize their selection methods to overcome that difficulty.[23]

Quantitative genetics also made it clear that Bakewell, an eighteenth-century practical breeder untrained in 'science,' had used the most effective 'scientific' selection method, even if his ability to statistically prove its effectiveness had been limited: namely, the progeny test. A technological factor that enhanced the capacity of quantitative genetics to utilize the progeny test effectively was AI or artificial insemination.[24] Renewed emphasis on progeny testing, better computers, and the use of AI made it possible to test males on an enormous scale. The now global, not North American, poultry-breeding industry catapulted forward in effectiveness.

Quantitative Genetics and Its Results

The productive result of fifty years' breeding work under these strategies and using this technology – that is, between 1950 and 2000 – was impressive. Layers averaged more than three hundred eggs a year, and broilers reached slaughter size earlier and on less feed. Studies comparing surviving mid-twentieth-century tester lines with modern commercial ones indicated that a great deal of the improvement seen in 2000 was due to breeding, not better feed or environment. The Canadian government had established an egg-laying tester line in 1950, a line that had been preserved to test against new strains. Hens at the end of the twentieth century could lay 344 eggs over the same period that the 'randombred' strain laid 267. Body weight of layers had been reduced by 20 per cent and the birds needed less feed.[25] In 2001 a similar study was done on the productivity of broilers by comparing modern broiler lines with a tester line. In 1957 the Canadian government established a tester strain for broilers, which had subsequently been maintained at the University of Georgia in Athens, and was known as the Athens-Canadian Randombred Control (ACRBC). A modern broiler line and the ACRBC were fed on both 1957 and 2001 feed rations to determine whether genetics or feed had brought about improvement. The results indicated that the commercial breeding companies (nature through genetic selection) were responsible for 85 to 90 per cent of the change. Nutrition (or nurture) accounted for only 10 to 15 per cent.[26] In spite of the significance of breeding for improvement, it might be noted that the cost of breeding represented only 10 to 15 per cent of the total cost of raising a broiler. Feed, housing, incubation, and so on accounts for the other 85 per cent. Feed alone amounts to between 60 and 70 per cent of total cost.[27]

International Corporate Breeding Culture

The broiler breeding industry has continued to be restricted in the United States by purchasing habits that had been established in the 1940s and 1950s by the Chicken-of-Tomorrow contests, which had awarded excellence on a divided sexual basis. An American marketing culture demanded that breeder companies only supply either the female line or the male of broilers, but not both, to the processor/integrator. (Single-sex selling had been used as a patenting device by egg-laying breeder companies when selling to breeders, not producers. Companies had been willing in the 1940s to sell genetics to other breeder organizations, and had protected themselves by marketing only one sex.) Integrators favoured this way of accessing broiler stock. They believed they maintained some independence as to judging the quality of the breeder's product in this fashion, and, as a result, directed how the broiler breeding companies would operate in the United States. In reality, the situation brought the integrator into the breeding process. Integrators made the decision about which male line would be crossed on which female line. In a generalized way, Americans believed that the female line provided the reproductive features in the cross, and the male gave a broiler its meat characteristics. The integrators argued that they were in the best position to judge their interests and therefore how to choose male and female lines to cross.

A single-sex selling culture did not take root in Britain or Europe. With the later development of breeder companies and no Chicken-of-Tomorrow campaign to encourage such a marketing strategy, British and European breeding operations that got under way in the 1950s (after recovery from war-damaged economies) looked at the problem of supplying genetics and protecting their interests differently. Their breeding philosophy emanated from the idea that the most effective heterosis in the end product, the commercial chick, could be had when the breeder selected for both the male and female lines that went into the chick. Breeding operations in Britain, such as the Ross outfit working out of Scotland with broiler stock, sold packages, not males and females separately. Heterosis in the hybrid chick ensured a biological lock, which in turn protected their interests. The purchaser could not reproduce hybrid chicks of the same quality by using commercial chicks in a breeding program.

British breeders of broiler genetics tended to look on American industry's way of buying stock as 'lottery genetics,' meaning any outcome was possible. By providing genetics as a combined package, breeders were able to test the nicking ability of various lines, something that an integrator could never do when purchasing males from one company and females from a different company. When a global situation with respect to sales developed for the breeding companies, this cultural pattern stayed in place, even though most geneticists in the world would argue by the 1990s that the European/British way was the more productive. In the United States, breeder companies are compelled to sell only a male or female line, making it difficult for the breeders to direct their breeding strategies. The same companies market packages everywhere else in the world. When British breeding lines entered the United States, through company mergers, they were forced to abandon European attitudes as to how sales worked. When the Ross operations became established in the United States, the company separated its male and female sales. The Ross male became a popular cross on the Arbor Acres female.[28] All breeder companies would prefer to sell packages in the United States. Crossing with some other company's genetics make it unclear which company's product is better and which inferior.[29] Breeders did not, as has been suggested by scholars, sell only one line to processors and producers in order to protect the intellectual property in the biology of the birds.[30] Marketing culture outside their control demanded that they market their products in that fashion. When breeders no longer sold to other breeder organizations, the selling of one sex for protection reasons no longer made sense and therefore could be abandoned. Heterosis from the combined lines, not the separation of sales by sex lines, provided the breeders with a biological lock when it came to selling to processors/producers.

Breeder companies tried to protect the genetics that resulted from their research, but recognized that it was difficult to do so. They seemed more relaxed, or perhaps one should say resigned, about the possibility of potential failure to preserve breeder secrets than they had been in the 1960s and 1970s.[31] It was hard to have complete control over the outflow of parent lines, a situation that promoted leakage and stealing. The excellent Ross male, for example, was probably stolen from the company that owns the line, and exists now within other competing breeding companies. Poultry geneticists, as a result, were inclined to call the breeding industry an 'industry of thieves.' Stealing might partially result from the fact that it was difficult to come by particular lines in any legitimate way. Breeder companies might wish to acquire certain genetics from other companies, but the only way to do so was to buy an entire breeder company. Individual lines were never sold.[32] Breeder trade in genetics rests solely on the buying and selling of companies.

Position of Fancy Breeders within the Modern Corporate Breeding World

With the infusion of genetics into utility breeding in North America, the fancy poultry industry went its own separate way, following the craft practices of traditional breeders, and working also within an international sphere with other fancy breeders. Cockerel breeding and pullet breeding as separate lines continued to prevail in many cases, because the Standard persisted in calling for perfection in each sex of some varieties that is incompatible with patterns of sex-linked inheritance. The breeding of lines in this fashion still attracts breeders because the method requires skill. The historic pattern of the breeder being separated from the poultry keeper has, apparently, continued to prevail as much in the fancy poultry industry as in the commercial industry. Many fanciers do not undertake serious breeding. Most of the people who make up the fancy industry and are owners of exhibition poultry buy their stock from the limited number of expert breeders.[33]

In recent years, a leading British fancy breeder and also a past president of the Poultry Club of Great Britain, W.C. Carefoot, set out how the modern fancier applies breeding principles. The writer claimed these did not reflect genetics. I think it is safe to argue that every principle below applies to the breeding methods used by geneticists and to breeding selection methods learned through experience. The poultry geneticist R.D. Crawford found striking parallels between Carefoot's approach to breeding and that of the industrial breeders.[34]

1 Hatch plenty of chicks from mated pens.
2 Select at all times for vigour, temperament, and other important traits.
3 Select the highest-quality birds for breeding, being careful not to breed in faults.
4 Line-breed to the extent of obtaining a considerable number of chicks from an outstanding bird, mating these so as to preserve both a high degree of quality and a high degree of relationship to the original superstar in the second generation, but then select the outstanding offspring and line-breed to these, not to the original.
5 Recognize the relationship between prepotency and homozygosity for the desired genes, and therefore consider the disqualification as future breeders of the stock which contains only a small proportion of good birds, since any good bird whose sibs are poor is highly unlikely to be prepotent; that is, use the progeny test.
6 Endeavour to ensure that in the selection processes, all desired virtues are evident somewhere in the strain so that they can in the future be blended into a perfect bird, remembering the 'like begets like' principle.

Together with the ability to correctly assess the quality of stock, the application of these six points is the key to successful breeding. The knowledge of which point to apply in a particular circumstance comes from experience. In fact, top-class breeders automatically assess their stock for breeding potential by combining visual pointers with known ancestry to predict the likely outcome of a particular mating. Such breeders are attempting to assess the genes carried unseen within the flock so as to control the results.[35]

A Canadian Corporate Breeder and His Success, 1940s–1980s: The Work of Donald McQueen Shaver

One Canadian breeder, Donald McQueen Shaver, developed a worldwide commercial breeding company in these changing times. Shaver Poultry Breeding Farms, developed from the late 1950s to the mid-1980s, is still part of the international corporate breeding structure.

Shaver was born in Galt, Ontario, in 1920. He began breeding at the age of twelve, and when fifteen he won the Canadian National Egg Laying contest. By 1936 he was running his own breeder/hatchery, named Grand Valley Breeders, and he sold day-old chicks over a radius of thirty-five miles. In 1940 Shaver went overseas to fight in the Second World War, and sent his flock to a breeder in Long Island, New York, for care. All were lost in a fire in 1944, and when Shaver returned from the war, he had to start all over again.[36]

It seemed to him that huge business opportunities existed in war-torn Europe and that there would be a demand for breeding material to produce food. From the beginning of his breeding endeavours as an adult, then, he was focused on international markets. He also came, as did other poultrymen, to see his work as part of a move to feed the millions of the world at reasonable prices. In 1946 he opened a hatchery and supported his breeding efforts by becoming a feed dealer for Maple Leaf Milling, Quaker Oats, Ful-o-Pep, and Ralston-Purina. He ran a feed store in downtown Galt.[37] Shaver bought the best stock he could find when he rebuilt his table-egg-laying breeding flock. His most successful female strains of Leghorns came from the Mount Hope operations, and what would be his best male Leghorn line was bought from a breeder in Texas (who was put out of the poultry business by disease, although Shaver never experienced that problem with the strain).[38] In 1954 these lines, when crossed, produced a Leghorn hybrid that he would soon market under the catchy name of Starcross 288. In 1985, when Shaver retired from the company he and the Starcross 288 had built, the company's press release argued that these hybrid layers produced one-third of the world's table white eggs.[39] The records of Shaver's operations offer an illuminating window into the world of poultry breeding in the critical years of breeder-company consolidation, geneticist breeder input, international markets, breeder secrecy and protection of intellectual property, and the shifting issues that affected significant breeding operations between the late 1950s and the mid-1980s.

In the late 1940s, Shaver worked within the Record of Performance structure for egg layers, and company papers reveal that even in the mid-1950s chicks from Shaver hatcheries tended to emanate from ROP-sired stock. He also produced chicks for the broiler industry, maintain-

Table 1 The Golden Cross that produced the Starcross 288. Strain codes from 1969, which show the 6 by 6 test that had identified the crossing of line 5 males on line 6 females, which was the original Starcross 288. (Shaver Papers, Archives, University of Guelph, series 1, box 8, file 13)

	Sire #1	Sire #2	Sire #3	STRAIN CODES Sire #4	Sire #5	Sire #6	Mean
Dam #1	207	237	228	232	242	243	232
Dam #2	231	230	232	245	251	249	240
Dam #3	210	225	197	220	242	231	221
Dam #4	220	244	227	217	235	237	230
Dam #5	238	242	238	223	239	242	237
Dam #6	252	246	241	244	**266**	224	246
Mean	226	237	227	230	246 Control	238	207

ing eight hundred pedigreed Barred Plymouth Rock females in his breeding flock and ultimately using them in cross-breeding programs. Shaver predicted, though, that the Barred Plymouth Rock was on the way out.[40] He was correct in that prediction. The White Plymouth Rock became a mainstay of commercial broiler-poultry production, due to its skin colour and the paler feather-pinhole markings left on that skin.

Shaver decided to break from the ROP in his egg-laying breeding. He believed he needed a hybrid to survive in the changing breeding world. Shaver was well aware of the effects of heterosis when he abandoned the ROP. His early breeding for egg-laying competitions before the war had shown him how powerful breeding for heterosis could be, and his original Starcross 288 confirmed the fact that the strategy worked. Shaver began to study material on genetics from Cornell, and he read Shull's classic 1908 work.[41] He became acquainted with Randall K. Cole, a professor at Cornell, who would play a role in the breeding programs undertaken by Shaver over the 1960s and 1970s that resulted in the further improvement of the Starcross 288. Some of Cole's advice reflected pure Felch theory on how to retain the genetic variation, or heterozygosis of original crosses, by crossing back to grandparents or even great-grandparents.[42] Cole devised a complicated breeding card-reference system, which allowed for the measurement of comparative crosses of lines or strains, and sub-strains within these.[43]

Shaver himself kept track of competing theories on how to breed for heterosis. He remained very sceptical of RRS and an intense inbreeding of lines. He preferred to work slowly for improvement by some form of family selection, generally by progeny or sibling. He kept notes, which he called 'breeding plans.' Some of his jottings in 1959 give a sense of how he believed breeding should proceed and why. Protection of the genetics in his breeding was paramount. 'Must protect self in sale of stock that could be used for breeding – hence – utilize multiple crosses of known pure stocks,' he wrote. A simple crossing of two lines did not provide adequate protection, it seemed, a factor that was just as important as good nicking between the lines. He elaborated:

> Mass reproduction of commercial chicks in hands of hatchery people who demand better production and reproduction than is often attainable in the 2-way cross, hence the use of the progeny of a two way cross as the female provides many advantages: a] To breeder: does not release pure lines to co-operators. b] To hatcheryman: more chicks per breeder are produced. Consequently, a minimum of three pure lines must be maintained. Rarely will three lines taken from among those available [work], hence need to test *many* lines in all combinations to spot those few which have commercial prospects.[44]

Perhaps ironically, after over ten years of geneticist input for breeding and testing to find a cross superior to the Starcross 288, none had been found. As Cole wrote to Shaver in 1968, 'Consequently one can say that all the effort and expense produced nothing of value. There is a good question as to whether continued effort in this direction is warranted.'[45] Apparently a good deal of luck, which any untrained breeder could experience with or without the aid of geneticist advice, went into finding a golden cross. Shaver always believed that to be so, and he questioned how significant factors of inheritance actually were to rates of improvement in stock. For egg production he believed that genetics played a 20 per cent factor and 80 per cent resulted from environment; 35 per cent of early sexual maturity came from the genetic background and 65 per cent from environmental factors; survival rates he set at 5 per cent genetics and 95 per cent environmental; mature body weight at 45 per cent genetics and 55 per cent environment; and, finally, feed conversion of egg layers was 14 per cent genetics and 85 per cent environment.[46] Clearly, in the nature-versus-nurture argument, Shaver favoured the influence of nurture. Genetics was the weak player in the game of producing good stock. Shaver's assessment of genetics in improvement, however, did not mean he dismissed the role of geneticists in breeding. He was one of the few Canadian breeders in the 1950s and 1960s to actively seek geneticist advice in breeding.

The Shaver collection reveals how important the role of evolving vaccines could be in the success of a breeder company's stock on the market. Better vaccines did not bring only sunshine to either breeders or producers. The Starcross 288 was particularly resistant to the worldwide disease known as Marek's. In the early years that was an advantage to Shaver, with respect to sales, especially internationally. With the advent of a good vaccine, however, the Starcross 288 lost some of its advantages, and R. Gowe, a geneticist with the Canadian department of agriculture, explained the situation to Shaver.

> You concentrated for years on breeding your parent lines under the home farm 'dirty' conditions. That meant that the Marek's virus was almost always present to some degree. Some or I guess most of the parent lines either had a high frequency of genes for resistance to Marek's or the frequency of the genes was increased under this exposure situation. You didn't use families that had high mortality and gradually increased the frequency of genes for Marek's resistance. Since Marek's was endemic throughout the world – the 288 which also was a strong nick for egg production and was constantly being improved – no doubt did generally better than competitors that didn't select for Marek's resistance as well as egg production etc. ... We now have good Marek's vaccines everywhere ... With Marek's vaccine the 288 advantage is reduced.[47]

Gowe did not advocate the breeding of stock non-resistant to disease because of advancing knowledge on vaccines. The danger was, however, that grower/producers would rely on vaccines to such a degree that care of housing and sanitary conditions would be considered less important than they were at a time when no vaccines were available. Since the birds could be vaccinated, it would be argued by some breeder companies that selective breeding should be concentrated on improvement of other factors than disease resistance, and that grower/producers would then concern themselves less with sanitation. Vaccines, then, could shift the emphasis put on certain breeding trends and practices followed in the industry.[48]

Breeder companies might try to keep their affairs and breeding strategies secret from the competitors, but the Shaver papers reveal that both breeders and academic geneticists were more aware of the internal environment of competitor companies than meets the eye. R.K. Cole reported to Shaver in 1974 on a significant poultry-breeding operation, Arbor Acres, which had achieved success and fame through the Chicken-of-Tomorrow campaigns of the late 1940s and early 1950s. Cole's comments are revealing about how the dynamics of major breeding companies could mean success or failure in a competitive world.

> Basically – Arbor Acres [is] on the skids ... The business aspects of AA [have] faltered – too many changes and perhaps a lack of understanding of the poultry industry ... A lack of adequate testing of their products before marketing them, plus unsupported claims of performance and a lack of appropriate service. I gathered that in some quarters feed salesmen knew better how to handle the AA birds than did the people from AA, who apparently were little concerned once the sale had been completed ... Their products have faltered in performance.[49]

When Shaver began selling in Europe he was forced to introduce new egg-laying lines due to the popularity of brown eggs in that region. In 1965 he brought out two new Starcross models, named the 555B and the 555R.[50] Both were based primarily on various crosses of Rhode Island Red strains. (By 2007 it was estimated that the brown-egg hen came mainly from a Rhode Island Red male and a White Plymouth Rock female. Smaller brown eggs resulted from a reversed sex cross of these two.)[51] The birds served a market in England, France, Belgium, Italy, and Portugal where brown eggs sold at a rate of 98 to 100 per cent versus white. Spain's market was about 50 per cent each. (In North America the market share of brown eggs to total was only 5 per cent by the 1980s.)[52] Shaver captured some of the European broiler market as well, with various crosses of White Plymouth Rocks and lines in the Cornish breed. The Cornish provided the male lines for crossing and the White Rock served as the basis of the female line, indicating that Shaver followed the basic golden-cross strategy that the Chicken-of-Tomorrow campaign had promoted. Broiler stock was sold under the name of Shaver Starbro.[53] He found it virtually impossible to penetrate the broiler market in North America, where what he called an 'old boys' club' controlled all breeding, probably meaning the breeder companies that had evolved from the successful competition of their founders in the Chicken-of-Tomorrow contest.[54] The names of the winners in that competition dominate the names of today's broiler-breeding companies and genetic lines – Cobb-Vantress, for example, and Saglio's Arbor Acres.

While by the 1960s Shaver had branch planted in European countries, he found it difficult to control how breeding and experimenting went on in these subsidiary companies. In Britain, for example, Shaver

GB (Great Britain) claimed an independence from the parent company as to breeding expertise that Shaver would not accept.

> Our position with Shaver GB is that we hold a controlling interest and this Company must do what we say, not agree in part or as they see fit ... We give orders, not suggestions ... Shaver GB has the ridiculous notion (and their press releases keep repeating this) that Shaver 577, to give one example, is their own product. They were the ones who discovered it and their attitude is that they can now cock a snoot at Shaver Canada. Of course ... all of the various English lines that were taken in, were done so under my instruction, even to the Lines ... I don't mind local company pride, but when it's against the parent company, that takes on a rather macabre degree of ridiculousness.[55]

The broiler situation in North America by the 1970s, that is, excess supply relative to demand, reduced the need of the integrators for more stock, and therefore made it even more imperative that breeders capture markets outside North America. The integrators themselves began to go into breeding. Some of this resulted from the fact that the integrators had stocked supplies that they could use without going back to the breeder, at least immediately. Some integrators began experimenting in their own breeding programs as well. Shaver did not think that this trend would last. 'I doubt that any board of directors of an integrator would allow their management to use 100 percent of their own breeding stock,' Shaver commented to other breeder companies.[56] Experimenting with breeding lines was an expensive proposition for an integrator, and Shaver's prediction proved, generally speaking, to be the right one. It took at least five years of working with various lines to find those that crossed productively enough to be economically viable. The integrators soon decided not to put the time and money into establishing such lines. The breeders continued to function separately. All breeding lines would be created by them in the future, no matter how closely a breeder company might work with an integrator. It was clear, however, from this temporary reduction in the North American supply in the 1970s, that in the long run breeder companies would need more sales abroad in order to survive. The situation encouraged the continued contraction and consolidation of North American breeder companies, as they attempted to penetrate markets around the world.

In 1985 Shaver retired and sold his company (operating in 94 countries). 'The poultry industry has been good to me,' he stated at the time of his retirement. 'There are many reasons for this,' he added, 'but chief among them, I believe, is the fact that I've always liked chickens and had a natural "feel" for them.'[57] His comments indicate the importance of having an innate interest in the birds themselves for successful breeding. Good breeders appreciated chickens outside their ability to generate money. Shaver sold his company to the feed company Cargill. He had sat on the board of Cargill since 1965, and many assumed that company had financed Shaver's operations after that time. Such thinking was in line with what most saw as the basic trends in the poultry industry: namely, integration, large corporate control, and the tendency of operations to take on monopoly-like features. It made sense that a feed company would ally itself with a breeder company. As small Ontario breeding operations went under, it was also assumed that the only way to survive was to be part of some large international corporation.[58] In reality, Shaver was asked to go on Cargill's board because of his international business experience and his ability to deal with Iron Curtain countries. Cargill had no financial connections to Shaver Poultry Breeding Farms until Cargill bought Shaver's operations. Shaver had been aware of the accusation in the 1960s and spent ten years trying to clarify the situation. In the end, he simply let it go.[59] Shaver's experience indicated that probably not all feed-company connections to poultry operations were as intimate as many believed to be the case.

Perhaps one of the most interesting facets of Shaver's breeding enterprises can be found in his beef-cattle breeding efforts, which began in the 1950s. These reveal how well attempts to create biological locks via heterosis in another livestock species, in order to ensure breeder intellectual property protection or to create markets, worked. Shaver aimed himself at cattle producers, or what would be known as producer/growers in the poultry world, not at cattle breeders. Starting with breeds that were uncommon in North America, he created a synthetic line, or composite breed (which he named Shaver Blend) based on crosses of the Lincoln Red, North Devon Red, and Maine Anjou breeds (he later added the genetics of five other breeds). He intended that bulls of this composite breed be crossed on the different prevailing genetics in North American herds, namely, Holsteins, Angus, and Herefords, in the production of market beef stock which would exhibit heterosis. It appeared that he hoped to maintain markets for his bulls by instituting a biological lock, thereby following standard poultry-breeding culture of the time. (Shaver never sold cows. 'They are gold to us,' his son explained. The holding back of females provided a biological lock.)[60] Cattle producers did not flock to Shaver Blend bulls, but not because they opposed the idea of working with hybrid vigour. It might be argued that it was resistance to biological locks, which effectively would remove them from deciding what market stock they wanted to produce, that underlay the attitudes of beef-cattle producer/growers to Shaver Blend bulls. Producer/growers had to be prepared to accept biological locks before the breeder could use heterosis to protect the intellectual property invested in breeding.

Cattle producers had used bulls for hybrid-vigour purposes for centuries. Classic crosses for hybrid vigour resulted from the breeding of Shorthorn cows to Angus bulls or Shorthorns cows to Galloway bulls. The progeny from such crosses were known as 'blue roans.' Angus were also crossed on Herefords and the progeny were called 'black baldies' – cattle that were black in colour rather than Hereford-red, but with the white faces of the Hereford. By the late 1960s cattlemen favoured crossing the newly imported breeds like Charolais and Limosin on traditionally bred Angus/Herford 'black baldies.' This three-way cross was designed to increase hybrid vigour. All such crosses, however, relied on the continued existence of purebred breeds that the purebred breeders generated. Cattle breeders were linked with the cattle producers who used their stock in a way that chicken breeders were not with producer/growers. To some degree the very nature of the beast – namely, the relative expense of females, the length of time to raise stock, and the space needed to do so – dictated how the cattle breeders interacted with cattle producers.[61] Beef producers played a role in how market animals were bred, even if they had to rely on the purebred breeders for the basis of their operations.

Failure to attract cattlemen to Shaver Blend bulls led to the development of the Shaver Blend as a 'purebred' breed of cattle. By the 1990s both cows and bulls were sold to breeders working together under Shaver guidance. The cattle were pedigreed under a private registry with established regulations for entry, the Shaver Beef Blend International Registry. Shaver Blend breeders formed an association of their own which continued to emphasize the values of hybrid vigour that their bulls offered when used on commercial cows. 'Pedigreed Bulls are available only from D.McQ. Shaver Beef Breeding Farms or from Licensed Shaver Herd owners,' promotional literature stated. 'All Licensed herds of Shaver Beef blend have access to Shaver's breeding recommendations which ensure the most effective matings to maximize the benefits of heterosis,' cattlemen were told.[62]

Canada's Management Supply System and the Broiler Industry, 1970s to the Twenty-first Century

A unique situation for breeder companies arose by the late 1970s as a result of particular rulings under a supply management system in Canada. Pressure to adopt supply management had existed in the United States from the 1920s, surfacing several times before the 1960s, when a formal plan was proposed and subsequently defeated under the Kennedy administration. A policy was established by which a farmer committee could propose supply management for the product the group grew. If two-thirds of the farmers growing the crop agreed with the committee's proposals, the referendum was deemed successful. The control plan would become law and be mandatory for any farmer involved in that particular industry. In 1963 a supply-management referendum on wheat supply failed to carry.[63] No other efforts to introduce supply management surfaced after that time in the United States.

Supply management of poultry products in Canada resulted from chicken farmers putting pressure put on government to stabilize marketing conditions and, furthermore, to control, first, interprovincial glutting of eggs and poultry meat as shortages and oversupply affected regions; and, second, the flood of American products into the country.[64] The Farm Products Marketing Agencies Act, passed by the federal government in 1972, created a structure under which marketing agencies and quota systems for producers could legally be established.

In the late 1970s, broiler imports from the United States became a serious problem. The Canadian Chicken Marketing Agency, formed under the Farm Products Marketing Agencies Act in 1978 (and renamed the Chicken Farmers of Canada in 1998), attempted to control that situation by operating a national supply-management program for broilers and allocating quota levels for production.[65] Supply management in Canada used production quotas for farmers and producer marketing boards to regulate and stabilize both supply and the farm prices of broilers. Under the quota system, farmers purchased a permit to produce a specified quantity of chickens. Quotas were bought on a one-time fee. Legally, quotas were retained by the marketing boards, but could be bought and sold on the open market. To avoid speculation, all quotas had to be used within one year. With quota in hand, broiler chicken farmers were free to sell chickens to processors at prices negotiated by the Chicken Farmers of Canada, which also allocated quota by province.[66] Today, the board of the Chicken Farmers of Canada (comprising ten provincial directors and four others) determines how much chicken meat will be produced during a particular growing period, and the supply management system mandates that processors must purchase all chickens produced. Quota holders (producer/growers) own the birds. (But the producer/grower has no control over the genetics he buys. The processor decides what to buy from a breeder company, owns the hatcheries and sells day-old chicks to the producer/grower. Processors themselves hold very little in the way of quotas.[67]

Although quota systems and market control under supply management in Canada, particularly in relation to broilers, has little to do with breeding directly, the supply-management structure supported other regulations that did concern the breeders, namely, housing and feeding conditions for broilers. Nature (genetics) interacted with nurture (feed and housing) conditions, a pattern that breeders were well aware of. Breeders had to address a factor they described as 'genotype × environment interaction,' if their genetics were to be successful in the marketplace, especially when compared to the product of their competitors. The importance of 'G × E interactions' seems to have been formally recognized by geneticists as early as 1936.[68] Better environmental conditions simply mean better performance from genetics embedded in the birds. (The Starcross 288's resistance to Marek's disease reflects lucky but unconscious G × E interaction breeding.) In other words, breeder companies could not ignore the conditions under which their chicks would be reared. The chicks had to be bred to cope with various environments.

For twenty years the Chicken Farmers of Canada have tried to control the welfare of the birds. Whether these efforts arose from an interest in animal welfare, or because they were deemed to help bring in the best profits for farmers on a joint basis, is not clear. It would seem likely that both factors played a role. What *is* clear is the fact that, because of these regulations, breeder companies are inclined to believe that chickens in Canada showcase their work more effectively than occurs in many other countries of the world, including the United States.[69] Canadian chicken farmers follow what is called the Codes of Practice for the Care and Handling of Poultry, which has strict guidelines. These have been reviewed and revised over the years. The most recent revision, done in 2003, was conducted in cooperation with the Canadian Federation of Humane Societies, the Canadian Veterinary Medical Association, and partners within the industry. Chickens in Canada are raised in clean, well-ventilated barns, with a constant supply of feed and fresh water.

They are free to roam over large barn floors, which are cleaned out after every flock is marketed (eight weeks), meaning some five or six times a year. Farmers have no reason to overcrowd their barns in order to get good returns, because supply management guarantees them a satisfactory income. Some 88 per cent of feed must be grains and grain by-products, birds are not fed steroids or hormones (illegal since the 1960s), and, the Chicken Farmers of Canada claim, the antibiotics given to them are carefully monitored. Chickens meant for meat do not have their peaks trimmed. 'Canadian chicken farms are second to none in cleanliness and far ahead of most,' argue the Chicken Farmers of Ontario.[70]

While international breeder companies like the Canadian supply-management system, many processors are violently opposed to it. Broiler and egg producers in the United States chafed over the restrictions that supply management in Canada placed on the American export industries. 'Supply management is an anachronism in the world of free trade,' the *Watt Poultry* press announced in 2007, and stated:

> The wholesale prices for poultry meat and eggs in Canada are 30 to 50% higher than in the USA attesting to the fact that the benefits accruing to producers in Canada are borne by consumers ... The inherent efficiencies associated with the scale of production and integration in the USA are clearly absent in the poultry industries of our neighbors to the North ... Maintaining barriers against importation by establishing annual tariff rate quotas with punitive duties which artificially maintain profitability among the domestic independent and cooperative producers is contrary to NAFTA and the spirit of WTO ... The current federal government is apparently committed to maintaining the system of supply management. Observers state that the government would be unable to fund a payout to farmers to compensate for the loss of equity represented by production permits ... Clearly some concessions will have to be made since the ultimate outcome is self evident. What should be debated is the rate of change and the procedures and policies to be applied. The best aspects of Canadian production should be maintained but the steady adoption of corporate ownership, free market expansion and partnering with US integrators [should be encouraged].[71]

Canadian processors would agree with this sentiment. The quota system is a contentious one from their point of view. Processors of poultry meat consider it a hindrance to good business operations, because supply management artificially rigs the market. Processors are told what they will buy and how much they must sell, regardless of market demand or capacity to absorb. Farmers, however, are supposed to be guaranteed an income in a way that is not true in the fully controlled, integrative situation in the United States. It has been argued that broiler-chicken farmers in Canada receive about 30 per cent of the final value of their product, while their American counterparts are thought to get merely 5 per cent.[72]

Overview of Changes in the Breeding Industry

The poultry-breeding industry has, over the past fifty or sixty years, collapsed into ever fewer and larger international companies. Many of the subsidiaries of these large holding companies originated in North America, but the operations that own them tend to be centred in Europe. Venture-capital organizations have seen economic potential in acquiring breeding companies. The breeding industry obviously appeared to give them an opportunity to buy cheaply (relatively speaking), reorganize, and sell at a profit. Breeding systems within the companies have tended both to revert to the basics of older systems and to take on new characteristics which relate to better and broader quantifiable results. The trend to looking at groups only, and in light of complex characteristics, has increased and is probably largely made possible by computer quantification. The breeding of poultry has come a long way from Sebright's days, but a surprising amount of Sebright's theory and Bakewell's before him, remains embedded in modern systems.

Conclusion

The development of modern agriculture, with its global emphasis, standardization, and mechanization through technology is a complicated story with many facets to it. The infiltration of scientific knowledge, biological knowledge in particular, is central to that story. Genetics would be one of the important arms of biology to interact with agriculture as it evolved in North America, even if it is by no means the only one. It became fundamental to the success of breeding, which surely is the critical 'basement' of agricultural production in any country. Understanding how genetics affected agricultural production in any country, however, is difficult because the historical development of that science in particular, and biology more generally, is complicated.[1] The process of evolution, and Darwinism generally, competed with Mendelism for attention from scientists. It has been argued that many early geneticists, in both Europe and the United States, after 1900 saw their work on heredity as studies in evolution, not research into the process of breeding.[2] Regardless of their interest in evolution, the work of academic geneticists was of little value to practical breeders around the world for some considerable time. Much contemporary literature devoted specifically to the practical breeding of animals, between 1900 and as late as 1950, failed to make clear what, if any, connections existed between genetics and farm-animal production. The scholar today is often confronted with a blank when looking at contemporary sources on the subject. An interrelationship often seemed to have been both taken for granted and, at the same time, rejected by people at the time. Under these conditions, livestock farming in particular has often seemed resistant to innovation based on genetics. The non-existent references to this linkage in primary sources makes it difficult today for scholars to understand whether or not farm-animal breeding was impervious to knowledge from genetics, though most historians would accept the logic that one must have affected the other in some way.

Considerable attention has been paid to the way genetic research and geneticists themselves worked to improve plant breeding in different parts of the world. The story of the interface of global livestock breeding with genetics has not been addressed to the same degree, probably partially because the effect of genetics has been less powerful in all the large livestock species, and therefore has not resulted in the same type of farming revolution. One livestock industry in North America, poultry breeding, followed international plant breeding paths rather closely after corn-breeding methodology in the United States developed along those lines. Patterns in poultry breeding, as they evolved in North America, echoed innovations in American corn breeding rather closely. There are subtle differences, however, because the traditional structure of breeding itself is complicated in any livestock industry. 'Breeding' results from the interconnected work of various sections within a livestock industry – each section performing a separate task. Scientific innovation must affect all aspects of breeding within a livestock industry, or at least change how all of them work in relation to each other, before any real revolution can take place. This book has focused on how genetics interacted with poultry breeding in North America, and thereby brought about a revolution, not just in the way chickens were bred, but also in the very structure of the breeding industry, which ultimately came to reflect a worldwide pattern.

Craft-breeding methods used from 1850 until roughly 1950 in the commercial American and Canadian poultry industries evolved from strategies laid down in Europe by eighteenth- and early-nineteenth-century animal breeders like Robert Bakewell and Sir John Sebright. The late eighteenth and early nineteenth centuries also saw the rise of important European breeder approaches to breeding, all of which ultimately played a role in how genetics would interface with craft breeding in North America. Breeders of improved stock in Britain and Europe began to categorize their breeding in three different ways: as thoroughbred, purebred, or standardbred. The three different outlooks provided the framework for breeder organizational bodies set up to protect the breeders in North America. All three also laid the groundwork for what would be an important and on-going pattern. The organization of breeding around these theoretical stances eventually led to an increased inability to separate breeding methods from breeding structures reflecting the theoretical underpinnings. The thoroughbred, purebred, and standardbred approaches were all utilized in the organization and regulation of North American poultry breeding after the mid-nineteenth century, and clouded which breeding methodology traditional or craft breeders actually used. The systems described what to expect in the end result of breeding, not how to breed for that result. That situation made it easier for American geneticists to argue that craft breeding involved no empirical thinking.

The commercial poultry industry in North America began with an interest in breeding birds first for exhibition purposes and then for egg production. A system was established in Britain in 1866 to regulate exhibition standards for various new breeds developed over the nineteenth century for beauty and use. The British system was adopted in the 1870s by the American Poultry Association, an organization that served breeders in both the United States and Canada. Breeding methodology revolved around the ideas of John Sebright, which were more elaborately articulated by Americans such as I.K. Felch and H.H. Stoddard. Strategies developed by these men all reflected some effort at preserving the beneficial results of inbreeding without incurring the dangers that result from that breeding practice. However, the inbuilt concern with beauty in conjunction with egg production in particular, regardless of breeding method, introduced a conflict in North American poultry breeding that would not be resolved until well into the twentieth century. The fancier became the expert breeder for both exhibition and farm birds, creating a dichotomy which took years to overcome. The practice of breeding and the sport of showing by the fancier, North Americans believed, demonstrated a breeder's ability to breed to whatever characteristic was desired.[3] Beauty also seemed to be the way to achieve productivity. Beautiful horses and cattle have a sound conformation structurally, which means they do not go lame, and can function without difficulties for a longer time than animals that are not structurally sound. There was little reason to think the situation was different with poultry. The emphasis on beauty helped polarize breeders into two camps, those who denied beauty was part of productivity and those who argued that the greater the beauty the more useful the bird. This cleavage played a particularly important role in the rise of genetics in relation to chicken breeding in the United States.

After 1900, with the rise of Mendelism, American geneticists also focused on improving the breeding of chickens for egg production. Most approaches they adopted were based on classical breeding methods that had been used since the eighteenth century by European naturalists who attempted to understand how heredity worked by experimenting with plant breeding. These methods relied on both inbreeding and crossing of those inbred lines. Over the first years of the twentieth century, editors of the North American poultry press deduced correctly that there was nothing new about any of these concepts. Much of the work done by classical geneticists focused on the creation of formulas which made sense of systems that had been discussed by poultry producer/growers and practical breeders for at least one hundred years.

Modern breeding strategies practised in North America did not differ that much from older ones. Sebright had clearly grasped the importance of controlling inbreeding levels as early as 1809, a concept fundamental to Sewell Wright's inbreeding coefficient, developed in the 1920s and later becoming a cornerstone of agricultural genetics. Today, all geneticist chicken breeders around the world use systems similar to those developed by the North American craft breeders in the nineteenth century. The breeding of birds to their great-grandparents is often practised, thereby following Felch's system described in the 1870s.[4] H.H. Stoddard's methods of intense inbreeding followed by outcrossing to non-related stock are also important to modern breeding. Called cyclic inbreeding, the system requires one or more generations of close inbreeding such as full-sibling matings, followed by a cross to unrelated lines of other inbreds, and then a return to close inbreeding for several more generations.[5] Johannsen's proposal that the organism be seen as a whole from an inheritance point of view was well understood by breeders in North America and elsewhere, as was the idea of phenotype versus genotype. Breeders knew that what an animal looked like would not necessarily dictate its genetic worth in terms of its ability to improve the breed or line. The bombardment of information in the poultry press by geneticists aroused scorn among North American craft breeders, and subsequently the suspicion that there was nothing new in the geneticist methodological approach to selection in breeding.

The rise of hatcheries after 1900, and the production of day-old chicks which could be shipped a considerable distance altered and complicated the position of breeders within the North American poultry industry. Mechanization via incubators allowed for a revolution in structure and breeding approaches. Breeders attempting to reach farm markets either opened hatcheries themselves and sold day-old chicks – more common before 1920 – or else formed relationships with larger hatchery operations, which were operated by businessmen with no other connection to agriculture. Hatcheries, then, formed an intimate, but increasingly complicated, relationship with the breeders in the production of day-old chicks. The perceived importance of breeding for the industry increased, in spite of the complex way the now larger, business-related, and mechanized breeder sector of the industry had developed by 1930. Hatcherymen became the titular heads of the breeding arm in both the United States and Canada, because they directed buying from breeders, multiplying of the genetics by flock owners called 'multipliers,' and the dispersal of resulting stock to producer/growers. Multiplier flock owners obtained their genetics under hatchery supervision. At this point the move to industry integration seemed to emanate out of hatchery affairs, and thus back through multiplier flocks to the breeders themselves, as numerous articles throughout the period on hatcherymen meetings quoted in the farm press make clear. Many believed the entire welfare of the North American poultry industry lay in the hands of hatcherymen. What they had to say about actual breeding, then, was important. The North American hatchery industry played a critical role in extending the influence of craft breeding, because it supported the exhibition structure for the breeding of egg-laying birds and also endorsed government regulations guiding standards for good hens, regulations built within a structure invented in Canada for chickens.

In Canada, a move to 'certify' and standardize birds for their ability to lay eggs resulted in the establishment of the Record of Performance (ROP) by government. The idea behind the ROP, namely, government endorsement of traditional breeder approaches to breeding, played into an on-going pattern: the masking of what craft breeding actually meant because the focus was on the standard, not the process of attaining it. While the ROP's historic roots in standardbred and purebred thinking laid the groundwork for this tendency, government backing of such systems was important to the trend. The inability to see the ROP for what it was (namely, a regulatory system designed to standardize quality) made it virtually impossible to understand that it did not explain how breeding should proceed. While the ROP tended to emphasize breeding by individual worth, for example, it did not dictate that approach. This was the final blow for craft breeding's visibility. When hatcheries increasingly focused on the ROP, craft-breeding methodology became virtually hidden. Rules set to standardize a breeding product came to be seen as a system for breeding itself. This trend was stronger in Canada than in the United States. The fact that hatchery accreditation became attached to the ROP structure through hatchery approval policies in the United States, though, encouraged the spread of ROP regulations. The ROP was never, however, supported by the best utility breeders in that country. It is the strains these breeders developed that went on to serve the new 'genetic' breeding industry of the United States and Canada. Experimental poultry breeding would increasingly move out of the public sector into the private, as breeding became big business supported by corporations which supplied capital input for extensive research and relied on its results rather than government to dictate how to breed. It was a trend that worried some.[6] The historical link between the hatcheries and the breeders broke down in the late 1940s, with the rise of American breeder companies who used geneticists. The breeder companies franchised hatcheries to sell their stock.

Perhaps somewhat surprisingly in view of the eventual outcome for them, the producer/growers of Canada and the United States played a role in that transition. Their historical preference for using hybrid or crossbred stock, more prevalent before the rise of standardbred/purebred chicken breeding and ROP structures, led to a natural affinity with professional breeders. The professionalization of breeding in the end 'deskilled' the producer/growers. Today they have no choice in what type of birds they work with. There may still be breeders and growers, but each interacts with the other in a completely different way than had been true in the past. Breeders are scientists attached to large companies that supply all poultry genetics to the world. Growers are farmers who raise birds. Growers no longer take part in any breeding process: they simply contract to raise a standardized product that often they don't even own.

When the hatcheries no longer directed breeding in North America, the ROP collapsed. The fall of the ROP represented other changing dynamics in the poultry-breeding industry, however, beyond the rising hegemony of egg-layer breeder companies. The overwhelming emphasis on breeding for egg laying, which was what the ROP was built on, had begun to decline. Many Americans who might have focused on egg breeding now turned to breeding for meat. Breeding efforts were aimed specifically at better egg production more than meat up until about 1940. It was only after 1940 that chicken flocks were bred for meat purposes, in the United States in particular. Before that time, meat tended to be viewed as a secondary industry, a side-line to egg production. Cultural factors seemed to play a role in the story of purpose separation of breeding emphasis in North American poultry breeding. One was the position of women within the poultry industry. Specialized single-purpose breeding did not occur in the United States and Canada until the hegemony of women in the poultry industry had ended. The gendered nature of the early poultry industry encouraged the widening of the already present structural division between the true breeder and the grower of the breeder's stock. Breeders were traditionally men, while producers of eggs and meat who used the breeder's stock were typically women. Producers (women) tended to go back to breeders (men) for replacements. Interest in meat production led to the masculinization of the North American poultry industry, as increasingly men took over production. The move of genetics into the field of meat production, or broiler production, consolidated the role of genetics in American poultry breeding. Virtually no traditional breeders were left.

It was the broiler industry that brought true business integration to the North American poultry industry. After the decline of hatchery hegemony, American feed companies became central to that development in both the United States and Canada. Feed companies supplied farmers with both feed and the broiler baby chicks that would eat the feed, and therefore often found it made economic sense to own hatcheries and produce baby chicks themselves. The process of broiler integration, then, which resulted in this baby-chick production and sale connection, reduced hatchery influence over primary breeders. The story of broiler integration is the story of the rise of the feed dealers and American feed companies, and the decline of the breeder/hatchery hegemony within the production industry. Feed companies sold off their broiler interests to processors by the 1970s. Within this integrative environment, the breeder companies remained separate.

The rapid consolidation and contraction of breeding companies began in the 1960s, and by the 1970s the pattern accelerated, with North American breeding operations being acquired by drug companies centred in both North America and Europe. European breeding companies that had been established in the 1950s were also acquired by drug interests, and in that way became amalgamated with North American breeding concerns. It would seem that the market for vaccines in the industry played a role in this development. By the 1990s the North American and European breeding companies were largely controlled by European corporations. By the early twenty-first century venture-capital concerns became involved, shrinking the number of international breeding operations still further. One such British company, after buying British- and American-based breeding companies, consolidated them into a new company, Aviagen, and subsequently sold it to a large German holding company. The main variant from this pattern is Cobb-Vantress, a breeding firm composed of smaller American companies. It is now owned by the huge American integrating processor Tyson Foods. As breeding companies developed, they tended to focus on the breeding of either broilers or table-egg layers but not both. There were, of course, exceptions. The more general trend to specialized egg or broiler breeding has stayed in place over the years, in spite of a contraction of company numbers. Breeder holding companies often own various subsidiaries that individually breed either broilers or table-egg layers. The Erich Wesjohann Group of Germany, for example, owns the table-egg breeding company Hy-Line International, as well as the broiler breeding company Aviagen.

If traditional breeding selection strategies used on chickens in North America reflected knowledge surprisingly similar to those put forward by geneticists, what was so revolutionary about genetics in relation to farm breeding of chickens in the United States and Canada? One of the most significant changes that American genetics brought to the practical farm world was a basic theoretical shift in orientation to the problem of how best to breed for improvement – but it was a shift, it is important to remember, that did not involve a new understanding of what the outcome of certain breeding approaches would bring. It was a question of outlook. Should birds be bred to reproduce truly or bred to produce stock that exceeded qualities in parents but did not breed truly? This fundamental divide between whether to breed for lines that bred truly or to breed stock unusable for future breeding lay at the heart of the matter. The American genetic methodological approach favoured the latter, which found particular favour in the United States, where the hybrid-corn breeding culture had been especially important in shaping attitudes towards what constituted scientific agricultural breeding. While the North American craft breeders traditionally favoured the former, they had fully understood these selection methods, the nature of the divide itself, and what any breeding system would produce since well before the advent of Mendel. In fact, heterosis, that cornerstone of American 'genetic' chicken breeding, had been recognized by practical breeders as early as Roman times. It was commercial considerations that drove an acceptance of this shift in outlook and, contingently, the adoption of a breeding methodology that was deemed to be 'scientific.' It is interesting to note, though, that neither craft breeder nor geneticist understood the process of heterosis. Geneticists were prone to dismissing the work of craft breeders on the basis that they did not know why predictable results followed certain breeding practices, but the scientists were not much better when it came to understanding this phenomenon. Attempts by geneticists to explain heterosis have been ongoing from the early twentieth century until the present.

In the 1930s one theory postulated that, not only did crossing inbred lines bring a return of heterozygosis and possibly the reduction of lethal recessives, but more importantly a sort of boomerang effect resulted from that re-establishment of heterozygosis, in which dominant alleles became even more dominant. This theory would be defined as 'over-dominance' in the 1950s. Dominance, or overdominance, was thought to reflect interactions among alleles at the same locus (meaning the point where two alleles join to create the gene).[7] A second mechanism for heterosis, called 'epistasis,' was later suggested. Epistasis explained heterosis as a reflection of allele interactions at different loci, which entailed a more complex way of looking at the effects of heterozygosis on the phenomenon of heterosis.[8]

Work on epistasis, specialized as that might be, serves as an example of how genetic research can work across various biological fields of knowledge and at the same time shows how difficult it can be to make genetic theory apply directly to farm production. Epistasis, like so many other genetic theories, was closely linked to the study of evolution. It was believed that epistasis theory helped explain how natural selection worked. The linkage of epistasis to heterosis, however, made it important to agricultural genetics. In the mid-twentieth century Motoo Kimura, a Japanese geneticist, used mathematical calculations to change attitudes to theoretical population genetics on the basis of his work in epistasis, work which had been initiated within an agricultural college in Japan and continued under Jay Lush (the most important livestock geneticist, worldwide, of the time) at Iowa State University, followed by the completion of a PhD at Wisconsin.[9] Kimura, working with mutation at the DNA level and therefore with molecular genetics, proposed that epistasis must reflect the fact that most evolutionary changes that occur do not alter a species; such changes are described, therefore, as being neutral – that is, they have no effect. He established the neutral theory, which is accepted today by molecular evolutionists interested in how heredity directs speciation. It is believed that dominance and epistasis play a role in the hybrid vigour resulting from cross-breeding in livestock, but the knowledge that it can be explained by the neutral theory does not easily relate to better selection methods as applied to livestock breeding.[10]

Fundamentally, in spite of evidence to support the validity of both theories as the basis of heterosis or hybrid vigour, the phenomenon has remained something of a mystery. In the 1980s an American, F.D. Hutt, noted that 'hybrid vigour is still one of the great enigmas in the field of genetics.'[11] Not much has changed since that time.[12] As one scientist said in 2006, 'The genetic basis of heterosis has been debated for nearly a hundred years without any emerging consensus.'[13] In spite of the fact that geneticists could not fully explain what caused heterosis, they did not hesitate over the years to advocate the breeding of stock for heterosis. The entrenchment of the hybridizing tradition might be one explanation for why scientific breeding in the United States continued to look only for heterosis. If sufficient time had been spent in earlier years by American geneticists on efforts to create better lines of chickens that bred on truly over generations, there is evidence that such lines ultimately would have been as productive as crosses yielding hybrid vigour. Even though D.F. Jones and E.M. East were aware by 1919 that breeding within pure lines and not crossing for heterosis would clearly yield better results for corn in the long run, there would be no changing course in the future.[14]

The major tool that genetics brought to North American breeding was sophisticated quantification on a far larger scale than any craft breeder had been able to do. Chicken breeding lent itself naturally to such an approach. Not only are the birds cheap individually, they are also inexpensive, relatively speaking, to house and feed compared to other farm animals. Quantifying results, the scientific approach, allowed chicken breeders to test various selection methods and to pinpoint how certain features were inherited. It was the interest of American geneticists in quantification that was at the heart of the matter as far as innovation was concerned, although that would not be evident for some time. Earlier breeders, for practical reasons involving their use of small numbers of animals, found quantification difficult to do. From the time of Bakewell, however, the best breeders in Europe and North America had always recognized that animal populations should be seen as breeding groups as much as a composite of individuals. The progeny test was based on that theory, even if it was applied only to restricted numbers. In the eighteenth century Bakewell attempted to use the progeny test as widely as possible by seeing what his breeding males could achieve on other farms than his own.

Why was American genetic breeding, which relied on heterosis and quantitative genetics but used methods of selection set out by craft breeders, able to so outstrip the productivity of the craft breeders? Part of the answer to this question lies in the fact that geneticists did not single-handedly bring about such a revolution. The role of superior vaccines and nutrition, for example, which arose over the same period in North American poultry husbandry, should not be underestimated. Statistics over the past fifty years on laying capacity and the improving quality of broilers in the United States and Canada make clear that it was only after the advent of American geneticist breeder companies that real progress was made. That fact suggests that corporate involvement was as essential to advancements in chicken breeding in any country as genetic knowledge itself. Companies can afford to deal with a sufficient volume to make effective use of quantitative genetics. Because they work with huge numbers of birds, high accuracy rates in testing are possible. Breeding units in international poultry companies are composed of so many more individual birds than could ever be possible in, for example, a single cattle enterprise. Artificial insemination technology has allowed for better testing of males across herds in the large animals (most particularly in the dairy industry, where large AI companies throughout the world fund extensive testing of bulls).[15] In spite of corporate involvement in AI, however, the cattle-breeding situation remains fractured when compared to the chicken-breeding industry, with its enormous capacity to assess various families and strains within huge chicken-breeding units. If all the cattle in the world were bred by only several companies – and there is some evidence in Europe that the trend to contraction of numbers of breeders is gaining momentum – the results would be more comparable.

At the same time, one must keep in mind the fact that it is the biological lock that heterosis offers that made corporate investment in breeding both attractive and feasible. It was the inbuilt biological lock implied by breeding for heterosis that allowed for the rise of American breeder companies and the consolidation of 'breeding' as a corporate enterprise on a large scale. The strategy made it possible for a breeder company to control the spread of the genetics it engineered through extensive testing. Quantitative volume breeding does not demand an emphasis on heterosis breeding, but corporate involvement in quantitative volume breeding does require heterosis breeding in order to survive. Corporate breeding, North American or global, is, therefore, inseparable from both. It is the triangle of quantitative breeding in huge volume, heterosis, and corporate involvement that explains what constitutes genetic success in chicken breeding, first in the United States and then around the world.

The interface of genetics with farm attitudes to poultry breeding in

North America became invisible partially because genetics expropriated much of the methodology of the older systems and called these approaches innovations of science. In the case of the poultry-breeding industry, much of what constitutes practical breeding methods of selection has been redefined as 'genetics,' or has been metamorphosed into genetics. This development, perhaps ironically, actually clouds how and what genetics contributed to this livestock industry as it evolved in North America, and also why it was able to do so. The contributions of both craft breeding and the innovations to breeding that genetics has offered have, as a result, been poorly understood within the North American context. Studies of genetic innovation diffusion with traditional breeding of chickens in other countries might well reveal similar patterns. Craft breeding, like genetic breeding, did not proceed in hermetically sealed national spheres.

Notes

Introduction

1 M.A. Jull et al., 'The Poultry Industry,' in United States Department of Agriculture (USDA), *Yearbook*, 1924, 378.
2 D. Fitzgerald, *The Business of Breeding: Hybrid Corn in Illinois, 1890–1940* (Ithaca: Cornell University Press, 1990), 1–3.
3 J. Harwood, 'Introduction to the Special Issue on Biology and Agriculture,' *Journal of the History of Biology* 39 (2006): 237.
4 M.D. Rossiter, *The Emergence of Agricultural Science* (New Haven: Yale University Press, 1975).
5 M.E. Derry, *Ontario's Cattle Kingdom: Purebred Breeders and Their World, 1870–1920* (Toronto: University of Toronto Press, 2001), 6, 8–9.
6 There is a vast amount of literature on the subject of agricultural education, agricultural experimentation, and farmer reaction to both within the framework of many countries. For the early situation in Ontario, Canada, see D. Lawr, 'Development of Agricultural Education in Ontario, 1870–1910,' PhD thesis, University of Toronto, 1972, and A.M. Ross and T.A. Crowley, *The College on the Hill: A New History of the Ontario Agricultural College, 1874–1999* (Ontario Agricultural College Alumni Association, Toronto: Dundurn Press, 1999, 2nd ed.). See as well J. Harwood, *Technology's Dilemma: Agricultural Colleges between Science and Practice in Germany, 1860–1933* (New York: Peter Lang Publishing Group, 2005). A few good references from the large field of work done on the situation within the United States (and which relate in particular to the material within this book) are D.C. Smith, *The Maine Agricultural Experiment Station* (Orono: University of Maine, 1980); A.I. Marcus, *Agricultural Science and the Quest for Legitimacy: Farmers, Agricultural Colleges, and Experiment Stations,*

1870–1890 (Ames: Iowa University Press, 1985); and R. Scott, *The Reluctant Farmer: The Rise of Agricultural Extension to 1914* (Chicago: University of Illinois Press, 1970).
7 Harwood, *Technology's Dilemma*.
8 J. Harwood, *Styles of Scientific Thought: The German Genetics Community, 1900–1933* (Chicago: University of Chicago Press, 1992).
9 S. Castonguay, 'The Transformation of Agricultural Research in France: The Introduction of the American System,' *Minerva* 43 (2005): 265–87.
10 Examples of science and practice in plant breeding outside corn are P. Dreyer, *A Gardener Touched with Genius: The Life of Luther Burbank* (Los Angeles: University of California Press, 1985), and Harwood, *Technology's Dilemma*.
11 For corn, see Fitzgerald, *The Business of Breeding*, and her 'Farmers Deskilled: Hybrid Corn and Farmers' Work,' *Technology and Culture* 34 (1993): 324–43. For emphasis on corn, and for the general use of hybridizing as a biological lock, see J. Kloppenburg, Jr, *First the Seed: The Political Economy of Plant Technology, 1492–2000* (Cambridge: Cambridge University Press, 1988). See also D.B. Paul and B. Kimmelman, 'Mendel in America: Theory and Practice, 1900–1919,' in *The American Development of Biology*, ed. R. Rainger, K. Benson, and J. Maienchein (Philadelphia: University of Pennsylvania Press, 1988).
12 For a discussion of hybrid corn in relation to the study of innovation diffusion see E.M. Rogers, *Diffusion of Innovations*, 4th ed. (New York: The Free Press, 1995), 31–6, 53–5.
13 A few examples are B. Kimmelman, 'Mr. Blakeslee Builds His Dream House: Agricultural Institutions, Genetics, and Careers 1900–1945,' *Journal of the History of Biology* 39 (2006): 241–80; C. Bonneuil, 'Mendelism, Plant Breeding and Experimental Cultures: Agriculture and the Development of Genetics in France,' *Journal of the History of Biology* 39 (2006): 281–308; T. Wieland, 'Scientific Theory and Agricultural Practice: Plant Breeding in Germany from the Late 19th to the Early 20th Century,' *Journal of the History of Biology* 39 (2006): 309–43.
14 See K.J. Cooke, 'From Science to Practice, or Practice to Science? Chickens and Eggs in Raymond Pearl's Agricultural Breeding Research, 1907–1916,' *Isis* 88 (1997): 62–86.
15 B. Theunissen, 'Breeding without Mendelism: Theory and Practice of Dairy Cattle Breeding in the Netherlands, 1900–1950,' *Journal of the History of Biology* 41 (2008): 637–76; and 'The Holsteinization of the Friesians: Culture of Dairy Cattle Breeding in the Netherlands, 1945–1995,' *Isis*, forthcoming.
16 R.J. Wood and V. Orel, 'Scientific Breeding in Central Europe during the

Early Nineteenth Century: Background to Mendel's Later Work,' *Journal of the History of Biology* 38 (2005): 239–72.

17 See R.J. Wood and V. Orel, *Genetic Prehistory in Selective Breeding: A Prelude to Mendel* (Oxford: Oxford University Press, 2001); Orel and Wood, 'Early Development in Artificial Selection as a Background to Mendel's Research,' *History and Philosophy of the Life Sciences* 3 (1981): 145–70; and Orel and Wood, 'Scientific Animal Breeding in Moravia before and after the Discovery of Mendel's Theory,' *Quarterly Review of Biology* 75 (2000): 149–57. See also V. Orel, 'Selection Practice and Theory of Heredity in Moravia before Mendel,' *Folia Mendelianna* 12 (1977): 179–99.

18 See, for example, S. Wilmot, 'From "Public Service" to Artificial Insemination: Animal Breeding Science and Reproductive Research in Early 20th Century Britain,' *Studies in History and Philosophy of Biological and Biomedical Sciences* 38 (2007): 411–41, and 'Between the Farm and the Clinic: Agricultural and Reproductive Technology in the Twentieth Century,' *Studies in History and Philosophy of Biological and Biomedical Sciences* 38 (2007): 303–15; C. Grasseni, 'Managing Cows: An Ethnography of Breeding Practices and Uses of Reproductive Technology in Contemporary Dairy Farming in Lombardy (Italy),' *Studies in History and Philosophy of Biological and Biomedical Sciences*, 38 (2007): 488–510.

19 Harwood, 'Introduction to the Special Issue.'

20 G.E. Bugos, 'Intellectual Property Protection in the American Chicken-breeding Industry,' *Business History Review* 66 (1992): 127–68. For the patenting of agricultural plants and animals more generally, see D. Kevles, 'The Advent of Animal Patents: Innovation and Controversy in the Engineering and Ownership of Life,' in S. Newman and M. Rothschild, eds, *Intellectual Property Rights and Patenting in Animal Breeding and Genetics* (New York: CABI Publishing, 2002), 18–30; his 'Patents, Protections, and Privileges,' *Isis* 98 (2007): 323–31 and his 'Protections, Privileges, and Patents: Intellectual Property in Animals and Plants since the Late Eighteenth Century,' in *Con/Texts of Invention*, ed. M. Biagioli, P. Jaszi, and M. Woodmansee (Chicago: University of Chicago Press, 2008); and H. Ritvo, 'Possessing Mother Nature: Genetic Capital in Eighteenth-Century Britain,' in J. Brewer and S. Staves, eds, *Early Modern Conceptions of Capital* (London: Routledge, 1995), 413–26.

21 A valuable and (it seems to me) a largely underutilized primary source (at least by historians) on genetics and its international effect on livestock and plant breeding, before the advent of molecular genetics, is USDA, 'Better Plants and Animals,' books 1 and 2 of *Yearbook* of Agriculture, 1936 and 1937. Henry A. Wallace, Secretary of Agriculture, ordered the research

done to complete these two very comprehensive volumes. Wallace and his family were major pioneers in the breeding of both hybrid corn and hybrid inbred laying hens (Hy-Line layers). John Hardiman, animal geneticist and vice-president of research and development at Cobb-Vantress, Inc. (interview, 7 May 2007), was a valuable source on molecular genetics and genomics. See also K. Laughlin, 'The Evolution of Genetics, Breeding and Production,' Temperton Fellowship, Report 15, Harper Adams University College, UK, 2007, 22–3; G. Bulfield, 'Strategies for the Future,' *Poultry Science* 76 (1997): 1071–4, and 'Sequencing of the Chicken Genome: An Overview,' December 2004 article in *Nature*, reproduced at The Poultry Site, http://www.thepoultrysite.com/FeaturedArticle/FAtype.asp?AREA.

22 See C.P. Van Tassell et al., 'SNP Discovery and Allele Frequency Estimation by Deep Sequencing of Reduced Representation Libraries,' *Nature Methods* 5 (2008): 247–52.

23 There is a vast amount of literature in, among others, the *Journal of Dairy Science*, *Nature Methods*, the *Journal of Animal Breeding and Genetics* in 2008 and 2009 on genomic testing via SNP markers.

24 See Alberta Department of Agriculture, 'Practical Poultry Keeping,' Bulletin 2, 1911; and I.K. Felch, *The Breeding and Management of Poultry* (Hyde Park, NY: Norfolk County Press, 1877); and his *Poultry Culture: How to Raise, Mate and Judge Thoroughbred Fowls* (Chicago: Donohue, Henneberry, 1902).

Chapter 1: Historical Background

1 J. Eriksson et al., 'Identification of the *Yellow Skin* Gene Reveals a Hybrid Origin of the Domestic Chicken,' *PLoS Genetics* 4 (2008): 1–8.

2 G. Sawyer, *The Agribusiness Poultry Industry: A History of Its Development* (New York: Exposition Press, 1971), 16; E. Brown, *Races of Domestic Fowl* (London: Edward Arnold, 1906), 70; R.D. Crawford, ed., *Poultry Keeping and Genetics* (New York: Elsevier, 1990), 3, 9–12.

3 *American Poultry Journal*, August 1924, 837; E. Brown, *Races of Domestic Fowl* (London: Edward Arnold, 1906), 2–14, 25; R.D. Crawford, ed., *Poultry Keeping and Genetics* (New York: Elsevier, 1990), 14–15.

4 M.A. Jull et al., 'The Poultry Industry,' in USDA, *Yearbook*, 1924, 382.

5 Ibid.

6 Crawford, *Poultry Keeping and Genetics*, 13–14.

7 W.D. Termohlen, 'Past History and Future Developments,' *Poultry Science* 47 (1968): 8.

8 G. Scott, 'The Learning Revolution, Distance Learning and Its Application

to a Global Poultry Industry: A U.K. Perspective,' Temperton Fellowship Report 11, Harper Adams University College, UK, 2003, 9.
9 E. Brown, *Races of Domestic Fowl* (London: Edward Arnold, 1906), 34.
10 Ibid., 10.
11 J. Bailey and G. Culley, *General View of the Agriculture of Northumberland, Cumberland, and Westmoreland* (London: B. McMillan, 1805; repr. Newcastle upon Tyne: Frank Graham, 1972), 251.
12 Crawford, *Poultry Keeping and Genetics*, 15.
13 Quoted in E. Brown, *Races of Domestic Poultry* (London: Edward Arnold, 1906) 35.
14 Crawford, *Poultry Keeping and Genetics*, 16–17.
15 Jull et al., 'The Poultry Industry,' 383.
16 For information on Bakewell see R. Trow-Smith, *A History of British Livestock Husbandry, 1700–1900* (London: Routledge & Kegan Paul, 1959); H.C. Pawson, *Robert Bakewell: Pioneer Livestock Breeder* (London: Crosby Lockwood, 1957) and 'Some Agricultural History Salvaged,' *Agricultural History Review* 7 (1959): 6–13; R.J. Wood and V. Orel, *Genetic Prehistory in Selective Breeding: A Prelude to Mendel* (Oxford: Oxford University Press, 2001); R.J. Wood, 'Robert Bakewell (1725–1795), Pioneer Animal Breeder and His Influence on Charles Darwin,' *Folia Mendelianna* 8 (1973): 231–42; D. Wykes, 'Robert Bakewell (1725–1795) of Dishley: Farmer and Livestock Improver,' *Agricultural History Review* 52 (2004): 38–55; and M.E. Derry, *Horses in Society: A Story of Breeding and Marketing Culture, 1800–1920* (Toronto: University of Toronto Press, 2006).
17 John Sebright, *The Art of Improving the Breeds of Domestic Animals* (London: John Harding, 1809), 9.
18 Bakewell to Culley, 8 February 1787, in 'The Bakewell Letters,' H.C. Pawson, *Robert Bakewell: Pioneer Livestock Breeder* (London: Crosby Lockwood, 1957), 107.
19 Quoted in George Mingay, ed., *Arthur Young and His Times* (London: Macmillan Press Ltd, 1975), 77–8.
20 Pawson, *Robert Bakewell*, 60, 62, 69.
21 M.E. Derry, *Bred for Perfection: Shorthorn Cattle, Collies, and Arabian Horses since 1800* (Baltimore: Johns Hopkins University Press, 2003), 18–19; and *Horses in Society*, 9; E. Heath-Agnew, *A History of Hereford Cattle and Their Breeders* (London: Duckworth & Co. Lt., 1983), 34.
22 Bakewell to Culley, 15 December 1791, in 'The Bakewell Letters,' Pawson, *Robert Bakewell*, 164.
23 Culley letter, 19 May 1792, ibid., 9, 11–12.
24 Sebright, *The Art of Improving*, 7.

25 Ibid., 5.
26 Ibid., 17–18
27 Ibid., 11.
28 Ibid., 10.
29 *American Poultry Journal*, November 1910, 1421.
30 Sebright, *The Art of Improving*, 17.
31 R. Bradley, *The Country Gentleman and Farmer's Monthly Director* (London: D. Browne, 1732), 10–11.
32 Derry, *Horses in Society*, 4–6, 31–3, and *Bred for Perfection*, 4–5; N. Russell, *Like Engend'ring Like: Heredity and Animal Breeding in Early Modern England* (Cambridge: Cambridge University Press, 1986), 2–4, 97–9, 104; D. Goodall, *A History of Horse Breeding* (London: Robert Hall, 1977), 33, 145, 146, 149; Trow-Smith, *History of British Livestock Husbandry*, 159–60; B. Tozer, *The Horse in History* (London: Methuen, 1908), 203–4.
33 Derry, *Horses in Society*, 10–12; *Bred for Perfection*, 7, 17–28; *Ontario's Cattle Kingdom: Purebred Breeders and Their World, 1870–1920* (Toronto: University of Toronto Press, 2001), 39–44.
34 Lady Wentworth, *Thoroughbred Racing Stock* (New York: Charles Scribner's Sons, 1938), 60.
35 Derry, *Horses in Society*, 241–4.
36 See M.H. Harper, *Breeding of Farm Animals* (New York: Orange Judd Co., 1920), 169, 174.
37 *Breeder's Gazette*, 20 April 1882, 537–8; 12 January 1882, 162; 16 March 1882, 395; 8 April 1896, 272; 2 November 1898, 422–3. See also Harper, *Breeding of Farm Animals*, 138–9; Derry, *Horses in Society*, 35–8; P. Thurtle, 'Harnessing Heredity in Gilded Age America: Middle Class Mores and Industrial Breeding in a Cultural Context,' *Journal of the History of Biology* 35 (2002): 43–78.
38 See, for example, *Farmer's Advocate*, January 1876, 13; February 1876, 27; March 1876, 46; 8 December 1910, 1927–8; and *Farming World and Canadian Farm and Home*, 1 January 1906, 161.
39 In the 1870s the examination of agriculture at the Ontario Agricultural College was on the history of Shorthorns, and revolved around the methods of the Collings and also Bakewell. See Ontario Legislature, Sessional Paper (henceforth Ontario, SP), 13, 1875–6, 31–2; and Ontario, SP 12, 1877, 48.
40 *American Poultry Journal*, August 1924, 837 and October 1908, 681; E. Brown, *Races of Domestic Fowl* (London: Edward Arnold, 1906), 41.
41 E. Brown, *Races of Domestic Fowl*, 36.
42 Jull et al., 'The Poultry Industry,' 384.
43 G. Burnham, *The History of the Hen Fever* (San Diego: Frank E. Marcy, 1935; repr. of 1855 edition), 263–4.

44 Ibid., 264, 266, 267.
45 Ibid., 273–4.
46 L. Wright, *The Brahma Fowl: A Monograph* (London: Journal of Horticulture and Cottage Gardener, 1870), 19.

Chapter 2: From Barnyard Scavenger to Bird of Beauty and Use

1 Enormous change was potentially possible in the birds because of the species' huge genetic capacity for variation, as genomics of the twenty-first century has made clear. The genetic diversity among modern chicken breeds is about half that between humans and chimpanzees. Any attempt to alter chicken characteristics through selection programs since the time of domestication has, in fact, been aided by this fact. 'Sequencing of the Chicken Genome: An Overview,' December 2004 article in *Nature*. Reproduced on The Poultry Site, http://www.thepoultrysite.com/FeaturedArticle/FAtype.asp?AREA. See, as well, D.W. Burt, 'Chicken Genome: Current Status and Future Opportunities,' *Genome Research* 15 (2005): 1692–8.
2 *American Poultry Journal*, November 1912, 1539–40.
3 See, for example, ibid., April 1911, 754.
4 T.F. McGrew, 'American Breeds of Poultry,' in US Department of Agriculture, Bureau of Animal Industry, *Report*, 1901: 520, 532; *Canadian Poultry Review*, December 1877, 2; *American Poultry Journal*, November 1912, 1539–40 and December 1912, 1693, 1696; E. Brown, *Poultry Breeding and Production*, vol. 1 (London: Ernest Benn Ltd, 1929) 290–7; O.A. Hanke, ed., *American Poultry History, 1823–1973* (Madison, WI: American Poultry History Society, 1974), 34–8; E. Brown, *Races of Domestic Poultry* (London: Edward Arnold, 1906), 150–2.
5 *American Poultry Journal*, January 1907, 33–4.
6 E. Brown, *Races of Domestic Poultry*, 72, 73.
7 Ibid., 74.
8 Ibid., 60–7; Hanke, ed., *American Poultry History*, 34–8.
9 *Canadian Poultry Chronicle*, April 1871, 149–50 and May 1921, 567; G. Sawyer, *The Agribusiness Poultry Industry: A History of Its Development* (New York: Exposition Press, 1971), 18; Hanke, ed., *American Poultry History*, 35–6.
10 *Canadian Poultry Chronicle*, December 1870, 82.
11 Hanke, ed., *American Poultry History*, 107.
12 *American Poultry Journal*, November 1911, 1569–71.
13 See, for example, *Farmer's Advocate*, 2 September 1920, 1526.
14 A.W. Foley, 'Practical Poultry Keeping,' Alberta Department of Agriculture, Poultry Bulletin 2, 1911, 29, 30.

15 An interesting discussion of the effects of shows on livestock breeding can be found in E.A. Heaman, *The Inglorious Acts of Peace: Exhibitions in Canadian Society during the Nineteenth Century* (Toronto: University of Toronto Press, 1999). Is now
16 *Farmer's Advocate*, September 1876, 177.
17 *American Poultry Journal*, February 1925, 341.
18 E. Brown, *Races of Domestic Poultry*, 203, 204.
19 *American Poultry Journal*, April 1925, 502, 512.
20 M.E. Derry, *Bred for Perfection: Shorthorn Cattle, Collies, and Arabian Horses since 1800* (Baltimore: Johns Hopkins University Press, 2003), 51–4.
21 *Farmer's Advocate*, February 1869, 25.
22 *Canada Farmer*, 1 January 1868, 11.
23 *Farmer's Advocate*, February 1869, 25 and April 1869, 55.
24 *Canadian Poultry Chronicle*, July 1870, 3.
25 Ibid..
26 *Canada Farmer*, 15 May 1868, 151.
27 *Poultry Journal*, January 1907, 33.
28 Ibid., December 1924, 1119, 1168, 1170–1.
29 *Farmer's Advocate*, 11 May 1905, 703; *Poultry Journal*, 1925, 368.
30 See, for example, *Poultry Journal*, March 1907, 267; August 1908, 585; March 1909, 293; and March 1925, 368.
31 Quoted in *Farmer's Advocate*, June 1888, 178.
32 *Canadian Poultry Review*, June 1892, 86.
33 *Farmer's Advocate*, 1 March 1900, 128 and 15 March 1900, 160.
34 *Canadian Poultry Review*, January 1894, 22–3.
35 See, for example, *American Poultry Journal*, August 1925, 756.
36 See, for example, ibid., May 1911, 1006.
37 McGrew, 'American Breeds of Fowl,' 525, 531–2; *American Poultry Journal*, March 1909, 307, August 1913, 1135–7, and October 1923, 989, 1015.
38 See Reliable Poultry Journal, *The Plymouth Rocks, Barred, White and Buff* (Quincy, IL: Reliable Poultry Journal Publishing Co., 1899), 23–26; *American Poultry Journal*, April, 1907, 365–6 and April 1908, 330–1.
39 *American Poultry Journal*, April 1915, 676.
40 *The Plymouth Rocks*, 26; *American Poultry Journal*, January 1922, 56, 58.
41 *The Plymouth Rocks*, 31.
42 *American Poultry Journal*, December 1912, 1693.
43 Ibid., October 1915, 1229, 1243, and November 1915, 1315–16.
44 See, for example, ibid., April 1909, 426, 428, 430.
45 Ibid., July 1922, 737.
46 Ibid., May 1911, 989–94 and June 1911, 1094.

47 *The Plymouth Rocks*, 14.
48 Ibid., 15.
49 Ibid., 16.
50 Reliable Poultry Journal, *The Leghorns: Brown, White, Black, Buff and Duckwing* (Quincy, IL: Reliable Poultry Journal, 1901).
51 *American Poultry Journal*, February 1914, n.p.
52 Ibid., February, 1915, 299, 300.
53 Jull et al., 'The Poultry Industry.'
54 *O.A.C. Review* 22 (1910), 244.
55 Sawyer, *The Agribusiness Poultry Industry*, 24.
56 R. Horowitz, 'Making the Chicken of Tomorrow: Reworking Poultry as Commodities and as Creatures, 1945–1990,' in *Industrializing Organisms: Introducing Evolutionary History*, ed. S.R. Schrepfer and P. Scranton (London: Routledge, 2004), 216.
57 Canada Census, vol. 2 (1901), 202, 278.
58 Canada Census, vol. 4 (1921), xcvi; Jull et al., 'The Poultry Industry,' 389.
59 *Advocate*, 25 February 1904, 273; Jull et al., 'The Poultry Industry,' 386, 432, 433, 434–6, 437.
60 Ontario Agricultural College, Department of Poultry Husbandry (University of Guelph Archives), Record Book Miscellaneous, 1899–1909.
61 *Farmer's Advocate*, 9 February 1905, 193.
62 *American Poultry Journal*, August 1910, 976–7.
63 W.D. Termohlen, 'Past History and Future Developments,' *Poultry Science* 47 (1968), 12.
64 *American Poultry Journal*, August 1911, 1268. A particularly clear description of Felch's inbreeding strategies and his chart appeared in an agricultural circular published in 1911 in Alberta. Foley, 'Practical Poultry Keeping,' 73.
65 An example might be seen in *American Poultry Journal*, November 1908, 736.
66 Ibid., December 1922, 1112–13. See November 1906, 910 as well.
67 Ibid., January 1928, 63–4.
68 See I.K. Felch, *The Breeding and Management of Poultry* (Hyde Park, NY: Norfolk County Press, 1877), 47. See as well I.K. Felch, *Poultry Culture: How to Raise, Mate and Judge Thoroughbred Fowls* (Chicago: Donohue, Henneberry, 1902).
69 *American Poultry Journal*, April 1911, 749.
70 See, for example, ibid., May 1912, 971 and July 1912, 1165.
71 Ibid., December 1913, 1518, 1520–1, 1542.
72 Quoted from American sources in *Canada Farmer*, 15 January 1874, 29.

Quoted from American sources in *Canadian Poultry Review*, March 1879, 54–5. Many, if not most, articles in the Canadian poultry press over the late nineteenth century were American articles picked up from the American press. The situation annoyed many Canadian editors, who complained that Canadians had too little to say.
73 Quoted in *Farmer's Advocate*, December 1868, 185.
74 Ibid., 1 May 1902, 348.
75 *Farming*, 18 January 1898, 156.
76 See, for example, *Canadian Poultry Chronicle*, February 1871, 118, 119 and October 1981, 49, 50, 51.
77 Ontario Legislature, Sessional Paper (henceforth Ontario, SP) 1, 1872–3, app. F, 446.
78 *Farmer's Advocate*, 16 September 1895, 367.
79 *American Poultry Journal*, January 1908, 36.
80 Perhaps one of the best sources for the interrelationship between 'purity' and the effects of hybridizing can be found in H. Ritvo, *The Platypus and the Mermaid and Other Figments of the Classifying Imagination* (Cambridge: Harvard University Press, 1997), 113–20.
81 It should be remembered that while the breeders resisted inbreeding, they continued to work with a restricted gene pool that had been enforced by a purebred system. This restriction implies mild inbreeding.
82 USDA, 'Better Plants and Animals,' book 1, *Yearbook* of Agriculture, 1936, 837, 847, 853, 864.
83 Ibid., 832–3, 837, 841–2, 844, 847, 853, 855.
84 *Canadian Poultry Review*, March 1879, 54–5. The Canadian poultry press had a tendency to rely on articles written by Americans, a fact that annoyed some. See *Canadian Poultry Review*, September 1878, 155–6.
85 *Farmer's Advocate*, April 1870, 51.
86 *Canadian Poultry Review*, September 1879, 183.
87 *Farmer's Advocate*, July 1883, 214.
88 See, for example, ibid., 26 August 1920, 1482, 20 February 1930, 267, and 24 January 1953, 37; (Toronto) *Globe*, 6 January 1915, 9; Select Committee of the House of Commons into Agricultural Conditions, *Proceedings*, 1924 (Ottawa, Canada), vol. 1, 521.
89 See, for example, *Farmer's Advocate*, 9 September 1937, 521 and 24 January 1953, 37.
90 See, for example, ibid., 4 January 1906, 14 and 11 April 1912, 688–9.
91 Ibid., 1 March 1923, 312; *Poultry Journal*, January 1907, 48.
92 *Farmer's Advocate*, 1 August 1900, 438.
93 Ibid., 23 July 1925, 1074.

94 Ibid., June 1882, 154.
95 See, for example, ibid., July 1882, 181, 20 July 1901, 443, July 1887, 208–9, August 1885, 237, 1 April 1899, 178, and 5 April 1900, 182; *Farm and Dairy*, 1 February 1917, 104; and *Farming*, June 1896, 615.
96 The situation was the same in Canada and the United States, as the farm press reveals. See S. McMurry, *Transforming Rural Life: Dairying Families and Agricultural Change, 1820–1885* (Baltimore: Johns Hopkins University Press, 1994), 203.
97 See Lu Ann Jones, *Mama Learned Us to Work* (Chapel Hill: University of North Carolina Press, 2002), 91, 92.
98 *Farmer's Advocate*, April 1877, 85.
99 Quoted ibid., November 1883, 339.
100 *The Globe*, 6 January 1915, 9.
101 *Farmer's Advocate*, 11 April 1912, 688–9.
102 Ibid., 14 March 1912, 481.
103 Jones, *Mama Learned Us to Work*, 84–5.
104 The important early poultry breeder I.K. Felch made perhaps the earliest general attempt to assess the importance of the poultry industry in the American agricultural economy. The poultry industry, he claimed, was the most significant livestock industry in the United States by the 1870s. The American census of 1870 set the value of all cattle, swine, and sheep sold for slaughter or slaughtered at roughly $399 million. Relying heavily on the importance of home consumption, the fact that most people lived on farms, and that virtually all farms kept fowl, Felch calculated the value of the poultry industry to the economy. He saw $500 million as a conservative estimate for the worth of the poultry industry. While Felch might well have overestimated its value, his reasoning that the poultry industry played a significant and hidden role as early as 1870 seems sound. Felch, *Breeding and Management of Poultry*, 20, 22.
105 *Farmer's Advocate*, 25 February 1909, 289.
106 W.C. Funk, 'Value to Farm Families of Food, Fuel, and Use of House,' Bulletin 410, US Department of Agriculture, November 1916, 5, 6, 1, 26.
107 Jones, *Mama Learned Us to Work*, 90.
108 R.D. Crawford, ed., *Poultry Keeping and Genetics* (New York: Elsevier, 1990), 1001.
109 V. McCormick, 'Butter and Egg Business: Implications from the Records of a Nineteenth-Century Farm Wife,' *Ohio History* 100 (1991): 63 4. V. McCormick, *Farm Wife: A Self Portrait, 1886 1896* (Ames: Iowa State University Press, 1990), 92–5.
110 H.W. Hawthorne, 'A Five-Year Farm Management Survey in Palmer

Township, Washington County, Ohio, 1912–1916,' Bulletin 716, US Department of Agriculture, September 1916, 27, 29–31.
111 Jones, *Mama Learned Us to Work*, 70, 73–4, 79, 84.
112 *Farmer's Advocate*, 14 July 1949, 574.
113 Ibid., 14 January 1900, 40.
114 *Canadian Poultry Review*, May 1892, 78.
115 See, for example, *Canadian Poultry Review*, March 1879, 55, 56.
116 *Poultry Journal*, October 1908, 677.
117 Ibid., December 1910, 1432.

Chapter 3: The Development of Agricultural Genetics

1 M.W. Rossiter, *The Emergence of Agricultural Science: Justus Liebig and the Americans, 1840–1880* (New Haven: Yale University Press, 1975), xi–xiv; Alan Marcus, *Agricultural Science and the Quest for Legitimacy: Farmers, Agricultural Colleges, and Experiment Stations, 1870–1890* (Ames: Iowa State University Press, 1985), 18–19.
2 W.D. Termohlen, 'The History of Development of Poultry Departments in the State Colleges or Universities of the United States,' *Poultry Science* 46 (1967): 294.
3 Marcus, *Agricultural Science*, 61, 63.
4 A.M. Ross and T.A. Crowley, *The College on the Hill: A New History of the Ontario Agricultural College, 1874–1999* (Toronto: Ontario Agricultural College Alumni Association and Dundurn Press, 1999, 2nd ed.).
5 US Department of Agriculture, *Yearbook*, 1936, 'Better Plants and Animals,' book 1, 227–8.
6 There is a great deal of literature on the entangled relationship of plant breeding to scientific and genetic research and government support. See, for example, K.J. Cooke, 'Expertise, Book Farming, and Government Agriculture: The Origins of Agricultural Seed Certification in the United States,' *Agricultural History* 76 (2002): 524–45.
7 D. Fitzgerald, *Every Farm a Factory* (New Haven: Yale University Press, 2003), 4, 35, 36.
8 For a broader look at the process of biology, and styles of thinking in science, see J. Harwood, *Styles of Scientific Thought: The German Genetics Community, 1900–1933* (Chicago: University of Chicago Press, 1992).
9 An important American plant breeder who was well known in Europe in this period was Joseph Cooper (1759–1840).
10 C.J. Bajema, ed., *Artificial Selection and the Development of Evolutionary Theory* (Stroudsburg, PA: Hutchinson Ross Publishing, 1982) 3, 4, 11–13, 22–5.

11 R. Olby, 'Mendel No Mendelian?' *History of Science* 12 (1979): 57. For a different point of view, see S. Müller-Wille and V. Orel, 'From Linnaean Species to Mendelian Factors: Elements of Hybridism, 1751–1870,' *Annals of Science* 64 (2007): 171–215.
12 C. Tudge, *In Mendel's Footnotes: An Introduction to the Science and Technologies of Genes and Genetics from the Nineteenth Century to the Twenty-Second* (London: Jonathan Cape, 2000), 64–6. R.J. Wood and V. Orel, *Genetic Prehistory in Selective Breeding: A Prelude to Mendel* (Oxford: Oxford University Press, 2001), 7. V. Orel, 'Selection Practice and Theory of Heredity in Moravia before Mendel,' *Folia Mendelianna* 12 (1977): 179.
13 Wood and Orel, *Genetic Prehistory in Selective Breeding*, 215, 237.
14 Bajema, ed., *Artificial Selection*, 7, 50–9.
15 V. Orel and R.J. Wood, 'Scientific Animal Breeding in Moravia before and after the Rediscovery of Mendel's Theory,' *Quarterly Review of Biology* 75 (2000): 150.
16 D.B. Paul and B.A. Kimmelman, 'Mendel in America: Theory and Practice, 1900–1919' in *The American Development of Biology*, ed. R. Rainger, K. Benson, and J. Maienschein (Philadelphia: University of Pennsylvania Press, 1988), 289, 293.
17 USDA, 'Better Plants and Animals,' 469.
18 L. Carlson, 'Forging His Own Path: William Jasper Spillman and Progressive Era Breeding and Genetics,' *Agricultural History* 79 (2005): 50, 52, 53, 58.
19 See Sessional Papers of the Ontario Legislature from 1879 until the late 1880s, which report on all aspects of the Ontario Agricultural College.
20 L. Quirk, *Prof. William Richard Graham: Poultryman of the Century* (Guelph: University of Guelph, 2006), 15.
21 Termohlen, 'History of Development of Poultry Departments,' 295–7.
22 Agriculture Canada, *One Hundred Harvests: Research Branch Agriculture Canada, 1886–1986* (Ottawa: Agriculture Canada, 1986), 197.
23 W.E. Shaklee, 'Federal-Grant Funds and Poultry Breeding Research in the United States,' *World's Poultry Science* 29 (1972): 217; Termohlen, 'History of Development of Poultry Departments,' 298.
24 Termohlen, 'History of Development of Poultry Departments,' 302–3, 298.
25 N. Russell, *Like Engend'ring Like: Heredity and Animal Breeding in Early Modern England* (Cambridge: Cambridge University Press, 1986), 40–1.
26 Paul and Kimmelman, 'Mendel in America,' 289.
27 J.A. Secord, 'Nature's Fancy: Charles Darwin and the Breeding of Pigeons,' *Isis* 72 (1981): 166–71, 174, 177. For more on Darwin's interest in artificial selection see S.G. Alter, 'The Advantages of Obscurity: Charles Darwin's

Negative Inference from the Histories of Domestic Breeds,' *Annals of Science* 64 (2007): 235–50; R.A. Richards, 'Darwin and the Inefficacy of Artificial Selection,' *Studies in History and Philosophy of Science* 28 (1997): 75–97; S.G. Sterrett, 'Darwin's Analogy between Artificial and Natural Selection: How Does It Go?' *Studies in History and Philosophy of Biolgical and Biomedical Sciences* 33 (2002): 151–68; L.T. Evans, 'Darwin's Use of the Analogy between Artificial and Natural Selection,' *Journal of the History of Biology* 17 (1984): 113–40; J.F. Cornell, 'Analogy and Technology in Darwin's Vision of Nature,' *Journal of the History of Biology* 17 (1984): 303–44; H.-J. Rheinberger and P. McLaughlin, 'Darwin's Experimental Natural History,' *Journal of the History of Biology* 17 (1984): 345–68; and B. Theunissen, 'Darwin and His Pigeons: The Analogy between Artificial and Natural Selection Revisted,' *Journal of the History of Biology* 44 (2011); online as of October 2011, but not yet printed.

28 C. Darwin, *On the Origin of Species* (London: John Murray, 1859), 13. The literature on Darwin and his knowledge of how heredity works is extensive. See, for example, S.E. Kingsland, 'The Battling Botanist: Daniel Trembly MacDougal, Mutation Theory, and the Rise of Experimental Evolutionary Biology in America, 1900–1912,' *Isis* 82 (1991): 484–5; and R. Olby, *Origins of Mendelism*, 2nd ed. (Chicago: University of Chicago Press, 1985), 40–7.

29 L. Wright, *The Brahma Fowl: A Monograph* (London: Journal of Horticulture and Cottage Gardener, 1870), 32–3.

30 Paul and Kimmelman, 'Mendel in America,' 293.

31 Kingsland, 'The Battling Botanist,' 489–93.

32 P.J. Pauly, *Controlling Life: Jacques Loeb and the Engineering Ideal of Biology* (New York: Oxford University Press, 1987), 3, 4, 56. See as well J. Sapp, 'The Struggle for Authority in the Field of Heredity, 1900–1932: New Perspectives on the Rise of Genetics,' *Journal of the History of Biology* 16 (1983): 311–42; and P.J. Pauly, 'The Appearance of Academic Biology in Late Nineteenth-Century America,' *Journal of the History of Biology* 17 (1984): 369–97.

33 B. Theunissen, 'Breeding without Mendelism: Theory and Practice of Dairy Cattle Breeding in the Netherlands, 1900–1950,' *Journal of the History of Biology* 41 (2008): 658; and his 'Knowledge Is Power: Hugo de Vries on Science, Heredity and Social Progress,' *British Journal for the History of Science* 27 (1994): 292, 298, 300, 301, 304, 307.

34 For a review of the historical debate over Mendel's role in the development of genetics, see Jan Sapp, 'The Nine Lives of Gregor Mendel,' in *Experimental Inquiries: Historical, Philosophical and Social Studies of Experimentation in Science*, ed. H.E. Le Grand (Dordrecht: Kluwer Academic Publishers, 1990), 137–66. See also R. Olby, 'Mendel no Mendelian?'

35 Some debate arose by the 1940s as to whether Spillman should be credited,

because of his 1901 paper 'Quantitative Studies on the Transmission of Parental Characters to Hybrid Offspring,' with the independent discovery of the basic laws of inheritance. It is clearly evident that Spillman was unaware of Mendelism at that time. But while Spillman appreciated the process of segregation in the inheritance of physical attributes, he did not – as Mendel did – see how the process of recombination in inheritance worked. L.P.V. Johnson, 'Dr. W.J. Spillman's Discoveries in Genetics: An Evaluation of his Pre-Mendelian Experiments with Wheat,' *Journal of Heredity* 39 (1948): 247–54.

36 Theunissen, 'Knowledge Is Power,' 293, 301, 304.
37 USDA, 'Better Plants and Animals,' 469–73.
38 D.C. Warren, 'A Half Century of Advances in the Genetics and Breeding Improvement of Poultry,' *Poultry Science* 37 (1958): 3.
39 L.C. Dunn, *A Short History of Genetics: The Development of Some of the Main Lines of Thought, 1864–1939* (New York: McGraw-Hill, 1965), 66–8, 70.
40 A particularly good source for definitions of terms is I.M. Lerner, *The Genetic Basis of Selection* (New York: John Wiley & Sons, Inc., 1958).
41 G.H. Shull, 'What Is "Heterosis?"' *Genetics* 33 (1948): 439–40. See, as well, F.B. Churchhill, 'William Johannsen and the Genotype Concept,' *Journal of the History of Biology* 7 (1974): 5–30.
42 Oren Harman and M.R. Dietrich, eds, *Rebels, Mavericks, and Heretics in Biology* (New Haven: Yale University Press, 2008), 70.
43 I.M. Lerner's *The Genetic Basis for Selection* is perhaps the easiest source for various breeding approaches. In order to understand how complicated the situation could be, though, see E. Goto and A.W. Nordkog, 'Heterosis in Poultry: Estimation of Combining Ability Variance from Diallel Crosses on Inbred Lines of Fowl,' *Poultry Science* 38 (1959): 1381–8 ; A.W. Nordkog et al., 'Heterosis in Poultry: Crossbreds versus Top Crosses,' *Poultry Science* 38 (1959): 1372–80; and A.W. Nordskog, 'The Evolution of Animal Breeding Practices – Commercial and Experimental,' Fourteenth Annual National Poultry Breeders' Roundtable, 1965, 51–66.
44 Dunn, *A Short History of Genetics*, 125.
45 Shull, 'What Is "Heterosis?"' 440.
46 S. Wright, *Evolution and the Genetics of Populations*, vol. 3, *Experimental Results and Evolutionary Deductions* (Chicago: University of Chicago Press, 1977), 11, 29; and Shull, 'What Is "Heterosis?"' 440.
47 D. Fitzgerald, *The Business of Breeding: Hybrid Corn in Illinois, 1890–1940* (Ithaca: Cornell University Press, 1990), 39.
48 D.F. Jones, 'Dominance of Linked Factors as a Means of Accounting for Heterosis,' *Genetics* 2 (1917): 471.

49 Ibid., 477.
50 Fitzgerald, *The Business of Breeding,* 55; J.R. Kloppenburg, *First the Seed: The Political Economy of Plant Biotechnology, 1492–2000* (Cambridge: Cambridge University Press, 1988), 99.
51 Fitzgerald, *The Business of Breeding,* 64.
52 O.A. Hanke, ed., *American Poultry History, 1823–1973* (Madison, WI: American Poultry History Society, 1974), 259.
53 Warren, 'A Half Century of Advances,' 4–5.
54 F.B. Hutt, 'Whither Poultry Genetics?' *Poultry Science* 21 (1965): 53.
55 I.M. Lerner, 'L.C. Dunn (1893–1974) and Poultry Genetics: A Brief Memoir,' *Journal of Heredity* 65 (1974): 185–6.
56 L.C. Dunn, 'Poultry Genetics Up to Date,' *Journal of Heredity* 24 (1933): 198.
57 Warren, 'A Half Century of Advances,' 3–4.
58 L.C. Dunn, 'Genetics at the Anikowo Station: A Russian Animal Breeding Center That Has Been Developed during the Reconstruction Period,' *Journal of Heredity* 19 (1928): 281–4; J. Cain and I. Layland, 'The Situation in Genetics: Dunn's 1927 Russian Tour,' *The Mendel Newspaper*, n.s., 12 (2003); and J. Marie, 'The Situation in Genetics II: Dunn's 1927 European Tour,' *The Mendel Newspaper*, n.s., 13 (2004).
59 W.E. Castle, *Heredity in Relation to Evolution and Animal Breeding* (New York: D. Appleton and Co., 1911), 2, 4.
60 Ibid., 151–2.
61 Ibid., 2, 4.
62 K. Rader, *Making Mice: Standardizing Animals for American Biomedical Research, 1900–1955* (Princeton: Princeton University Press, 2004), 31.
63 See W.E. Castle, 'Pure Lines and Selection,' *Journal of Heredity* 5 (1914): 93–7.
64 See S. Wright, 'Mendelian Analysis of the Pure Bred Breeds of Livestock, Part 1: The Measurement of Inbreeding and Relationship,' *Journal of Heredity* 14 (1923): 339–48; and 'Mendelian Analysis of the Pure Bred Breeds of Livestock, Part 2: The Duchess Family of Shorthorns as Bred by Thomas Bates,' *Journal of Heredity* 14 (1923): 405–22.
65 For Wright's description of how his work related to agriculture, see S. Wright, 'The Relation of Livestock Breeding to Theories of Evolution,' *Journal of Animal Science* 46 (1978): 1192–1200.
66 See J.F. Crow, 'Sewell Wright's Place in Twentieth-Century Biology,' *Journal of the History of Biology* 23 (1990), 58, 62–6.
67 W. Provine, *Sewell Wright and Evolutionary Biology* (Chicago: University of Chicago Press, 1986), 156.
68 Provine, *Sewell Wright and Evolutionary Biology*, 138–9, 140, 141; and *The*

Origins of Theoretical Population Genetics (Chicago: University of Chicago Press, 1971), 160–1.
69 Quoted in Provine, *Sewell Wright and Evolutionary Biology*, 156. See, as well, S. Wright, 'The Effects of Inbreeding and Crossbreeding on Guinea Pigs,' Bulletin 1090, USDA, 1922; and *Systems of Mating and Other Papers* (Ames: Iowa State College Press, 1958), which contains reprints of 'Systems of Mating,' *Genetics* 6 (1921): 111–78, 'Evolution in Mendelian Populations,' *Genetics* 16 (1931): 97–159, 'Correlation and Causation,' *Journal of Agricultural Research* 20 (1921): 557–85, and 'The Method of Path Coefficients,' *Annals of Mathematical Statistics* 5 (1934): 161–215.
70 There are countless examples of this, among them E.B. Babcock and R.E. Clausen, *Genetics in Relation to Agriculture* (New York: McGraw-Hill Book Co., Inc., 1918); and C. Wriedt, *Heredity in Livestock* (London: MacMillan and Co., 1930).
71 Wright's own work in the *Breeder's Gazette* and through the Bureau of Animal Industry over the years on the subject was very difficult for farmers to follow.
72 For a rather good, short overview of population genetics see Crow, 'Sewell Wright's Place in Twentieth-Century Biology,' 68–73.
73 See A.R. Hallauer, 'History, Contribution, and Future of Quantitative Genetics in Plant Breeding: Lessons from Maize,' *Crop Science* 47 (2007), from *International Plant Breeding Symposium*, 2007, S4–19, for how population genetics related to quantitative genetics.
74 Lerner, *The Genetic Basis for Selection*, 271, 272.
75 In order to understand the basic differences between population genetics and quantitative genetics see, for example, E. Pollak et al., eds, *Proceedings of the International Conference on Quantitative Genetics* (Ames: Iowa State University Press, 1977); B.S. Weir et al., eds, *Proceedings of the Second International Conference on Quantitative Genetics* (Sunderland, MA: Sinauer Associates, Inc., 1988); F. Pirchner, *Population Genetics in Animal Breeding* (San Francisco: W.H. Freeman and Co., 1969); R.E. Comstock, *Quantitative Genetics with Special Reference to Plant and Animal Breeding* (Ames: Iowa State University Press, 1996); J.F. Crow, *Basic Concepts in Population, Quantitative, and Evolutionary Genetics* (New York: W.H. Freeman and Co., 1986); E.P. Cunningham, *Quantitative Genetic Theory and Livestock Improvement* (Hanover, NH: University Press of New England, 1979); and D.S. Falconer, *Introduction to Quantitative Genetics* (Edinburgh: Oliver and Boyd, 1960).
76 See Wright, 'Mendelian Analysis of the Pure Bred Breeds of Livestock.'
77 D.C. Warren,' Breeding' in Hanke, ed., *American Poultry History, 1823–1973*, 273.

78 G.B. Havenstein, 'Performance Changes in Livestock Following 50 Years of Genetic Selection,' *Lohmann Information* 41 (December 2006): 30. See also Theunissen, 'Breeding without Mendelism,' 637–76, for the effects of Lush on both Dutch geneticists and British quantitative geneticists.
79 Provine, *Sewell Wright and Evolutionary Biology*, 321–6.
80 See, for example, J. Lush, *Animal Breeding Plans* (Ames, IA: Collegiate Press, Inc., 1937).
81 J. Lush, 'Genetics and Animal Breeding,' in *Genetics in the 20th Century*, ed. L.C. Dunn (New York: MacMillan Co., 1951), 494.
82 Ibid., 496–9.
83 Crawford, ed., *Poultry Keeping and Genetics*, 956.
84 C.F. McClary, 'Reciprocal Recurrent Selection Response in Poultry, Other Animals and Plants,' Eighteenth Annual National Poultry Breeders' Roundtable, 1969, 121.
85 J. Lush, 'Improving One Character by Breeding for Another,' Fact Finding Conference of the Institute of American Poultry Industries, 1958–9, 186.
86 USDA, 'Better Plants and Animals,' 831, 863, 888, 891, 908–9.
87 C. Bonneuil, 'Mendelism, Plant Breeding and Experimental Cultures: Agriculture and the Development of Genetics in France,' *Journal of the History of Biology* 39 (2006): 294.
88 USDA, 'Better Plants and Animals,' 131.
89 See, for example, USDA, 'Better Plants and Animals,' 832.
90 See C. Strom, *Profiting from the Plains: The Great Northern Railway and Corporate Development of the American West* (Seattle: University of Washington Press, 2003), 129–31; M.E. Derry, *Ontario's Cattle Kingdom: Purebred Breeders and Their World, 1870–1920* (Toronto: University of Toronto Press, 2001), 102–3.
91 Warren, 'A Half Century of Advances,' 6.
92 Ibid., 14.
93 USDA, 'Better Plants and Animals,' 985–7.

Chapter 4: Breeding for Eggs in North America

1 D. Fitzgerald, *Every Farm a Factory* (New Haven: Yale University Press, 2003), 5.
2 Board of Inquiry into the Cost of Living, *Report*, vol. 1 (Ottawa, Canada, 1915), 782.
3 *Advocate*, 25 February 1904, 273; M.A. Jull et al., 'The Poultry Industry,' in USDA, *Yearbook*, 1924, 386, 432, 428, 433, 434–6, 437; Board of Inquiry into the Cost of Living, *Report*, vol. 1, 782, 784, and vol. 2, 1103, 1104;

Select Committee of the House of Commons into Agricultural Conditions, *Proceedings*, vol. 1 (Ottawa, Canada, 1924), 253; J.H. Hare and T.A. Benson, 'Farm Poultry and Egg Marketing Conditions in Ontario County,' Bulletin 208, Ontario Department of Agriculture, 1913, 1, 3, 17, 18, 19.
4 Lu Ann Jones, *Mama Learned Us to Work* (Chapel Hill: University of North Carolina Press, 2002), 57, 79, 86.
5 Board of Inquiry into the Cost of Living, *Report*, vol. 1, 782.
6 M.C. Urquhart and K.A. Buckley, *Historical Statistics of Canada* (Toronto: MacMillan Co. Ltd., 1965), 367.
7 *Advocate*, 29 April 1909, 672–3 and 9 December 1920, 2143; *Agricultural Gazette of Canada* 6 (1919): 731.
8 G. Sawyer, *The Agribusiness Poultry Industry: A History of Its Development* (New York: Exposition Press, 1971), 24.
9 O.A. Hanke, ed., *American Poultry History, 1823–1973* (Madison, WI: American Poultry History Society, 1974), 355.
10 W. Johannsen, 'The Genotype Conception of Heredity,' *American Naturalist* 45 (1911): 142–3.
11 J. Sapp, 'The Struggle for Authority in the Field of Heredity, 1900–1932: New Perspectives on the Rise of Genetics,' *Journal of the History of Biology* 16 (1983): 336–7, 324.
12 K. Cooke, 'From Science to Practice, or Practice to Science? Chickens and Eggs in Raymond Pearl's Agricultural Breeding Research, 1907–1916,' *Isis* 88 (1997): 67, 70–4, 76, 77, 82. See, as well, R. Pearl, 'Breeding Poultry for Egg Production,' Bulletin 192, in *Annual Report* of the Maine Agricultural Experiment Station, 1911; and 'Inheritance of Hatching Quality of Eggs in Poultry,' *American Breeders' Magazine* 1 (1913): 129–33; R.R. Slocum, 'Poultry Breeding,' *Journal of Heredity* 6 (1915): 484–6; Research Committee on Animal Breeding, 'Live-Stock Genetics,' *Journal of Heredity* 6 (1915): 21–2.
13 See, for example, Legislature of Ontario, Sessional Paper (henceforth Ontario, SP), 39 (1916), 40.
14 Jull et al., 'The Poultry Industry,' 410.
15 Slocum, 'Poultry Breeding,' 486.
16 *American Poultry Journal*, September 1913, 1186.
17 Maine Experiment Station, Bulletin 305 (1913), 388. This quote appeared in the *American Poultry Journal*, May 1913, 847.
18 Ibid., May 1913, 847.
19 Ibid., April 1913, 672.
20 Ibid., May 1913, 847.
21 Ibid.
22 See J. Marie, 'For Science, Love and Money: The Social Worlds of Poul-

try and Rabbit Breeding in Britain, 1900–1940,' *Social Studies of Science* 38 (2008): 925 for more on this phenomenon.
23 Crawford, *Poultry Breeding and Genetics*, 1024, 1026.
24 F.B. Hutt, 'Seventy-five Years of Poultry Genetics,' Roundtable of Poultry Breeders Association, 1975, at www.poultryscience.org/pba/1952-2003/1975/1975%20Hutt.pdf, 152.
25 *American Poultry Journal*, October 1913, 1278.
26 Ibid.
27 *American Poultry Journal*, December 1913, 1517.
28 Ibid., January 1921, 117.
29 Ibid., April 1922, 478.
30 Marie, 'For Science, Love and Money,' 922.
31 *American Poultry Journal*, February 1929, 148.
32 Ibid., December 1932, 17–18.
33 See, for example, *American Poultry Journal*, April 1907, 372–4; November 1908, 735–7; September 1910, 1042–3; and November 1910, 1267–8.
34 *American Poultry Journal*, April 1915, 693.
35 *Advocate*, 15 June 1903, 559.
36 *American Poultry Journal*, September 1907, 690.
37 Ibid., August 1910, 989.
38 *Advocate*, 12 January 1905, 50 and 6 November, 1919.
39 Ibid., 8 February 1912.
40 Ibid., 20 November 1920, 2035.
41 See ibid., 1 August 1912, 1369.
42 Ibid., 11 November 1920, 1954.
43 *American Poultry Journal*, November 1925, 938–9.
44 Ibid., April 1924, 515, 550.
45 Ibid., November 1915, 1321–2.
46 See, for example, D.K. Flock and R. Preisinger, 'Genetic Changes in Layer Breeding: Historical Trends and Future Prospects,' unpublished manuscript, ca. 1999, 24–7.
47 For examples see *American Poultry Journal*, February 1912, 298 and June 1921, 642.
48 See M.E. Derry, *Ontario's Cattle Kingdom: Purebred Breeders and Their World, 1870–1920* (Toronto: University of Toronto Press, 2001) and *Bred for Perfection: Shorthorn Cattle, Collies, and Arabian Horses since 1800* (Baltimore: Johns Hopkins University Press, 2003) for detailed discussions of the interrelationship of livestock (and dog) breed associations with government in both the United States and Canada over the late ninetenthth and early twentieth centuries.

49 Derry, *Ontario's Cattle Kingdom*, 73–83; and *Bred for Perfection*, 36–44.
50 G.E. Reaman, *History of the Holstein-Friesian Breed in Canada* (Toronto: Collins, 1946), 320, 324–5, 340, 341, 343.
51 E.S. Snyder, 'A History of the Poultry Science Department at the Ontario Agricultural College, 1894–1968' (unpublished manuscript, 1970), 92–3. See as well L. Quirk, *Prof. William Richard Graham: Poultryman of the Century* (Guelph: University of Guelph, 2005).
52 *Advocate*, 15 May 1919, 970.
53 Ibid., 7 August 1919, 1429–30; 'Official Record of Performance for Poultry,' *Agricultural Gazette of Canada*, 1919, 796.
54 *American Poultry Journal*, April 1926, 464.
55 M. Jull, 'A System for Pedigreeing Poultry,' *Agricultural Gazette of Canada*, 1923, 40, 41, 42–5.
56 *Advocate*, 29 March 1923, 469; *Canadian National Record Book*, 1932, 43, 45, 56–9, series 1, box 5, file 4, Shaver Collection, Archival and Special Collections, University of Guelph Library.
57 *Advocate*, 10 March 1938, 153.
58 Ibid., 27 October 1927, 1561.
59 See *American Poultry Journal*, August 1922, 806 and January 1925, 19, 77, 78–81.
60 Ibid., May 1923, 543.
61 See ibid., June 1921, 642 and June 1922, 672–3.
62 Ibid., April 1926, 466, 468.
63 Ibid., June 1922, 674.
64 Ibid., April 1926, 466, 468.
65 Ibid., June 1924, 711, 730.
66 Hanke, ed., *American Poultry History*, 703, 704.
67 See Derry, *Ontario's Cattle Kingdom*, *Bred for Perfection*, and *Horses in Society* for a considerable amount of discussion involving cattle and horse-breed association frictions internationally, as well as the relationship of breed associations to government.
68 A.L. Hagedoorn and G. Skyes, *Poultry Breeding: Theory and Practice* (London: Crosby Lockwood & Son Ltd, 1953), 217–20; Hanke, ed., *American Poultry History*, 702, 703, 704.
69 *American Poultry Journal*, August 1923, 893–4.
70 D.C. Warren, 'A Half Century of Advances in the Genetics and Breeding Improvement of Poultry,' *Poultry Science* 37 (1958): 13.
71 R.D. Crawford, ed., *Poultry Keeping and Genetics* (New York: Elsevier, 1990), 989.
72 Warren, 'A Half Century of Advances,' 13.

73 USDA, 'Better Plants and Animals,' part 1, *Yearbook* of Agriculture, 1936, 874–5, 876–9, 898–9, 904.
74 *American Poultry Journal*, January 1927, 11, 88, 90–2, 94–7. See as well Fitzgerald, *Every Farm a Factory*, 106, 115.
75 *American Poultry Journal*, February 1927, 143.
76 Ibid., December 1925, 1032, 1038.
77 Hanke, ed., *American Poultry History*, 253.
78 Walker van Riper, 'Aesthetic Notions in Animal Breeding,' *Quarterly Review of Biology* 7 (1932): 84–7; book review of *American Dairy Cattle: Their Past and Future* in *Journal of Economic History* 2 (1942): 227–9.
79 E. Parmelee Prentice, 'Food for Americans, 1980– : The Supply of Animal Proteins – The Agricultural Colleges,' *Political Science Quarterly* 66 (1951): 483. Hanke, ed., *American Poultry History*, 260.
80 Van Riper, 'Aesthetic Notions in Animal Breeding,' 84–7; book review on *American Dairy Cattle*.
81 Hanke, ed., *American Poultry History*, 254–5, 269, 704.
82 *American Poultry Journal*, August 1923, 881.
83 Ibid., June 1923, 749.
84 Ibid., September 1924, 902–3.
85 Ibid., October 1924, 990, 991.
86 A.L. Hagedoorn, *Animal Breeding* (London: Crosby Lockwood & Son Ltd, 1939; 6th ed. 1962), 10.
87 University of California, 'In Memoriam, September 1978 – I. Michael Lerner, Genetics: Berkeley,' http://content.cdlib.org/view?docId =hb4q2nb2nd&chunk.id=div00043. R.W. Allard, 'Israel Michael Lerner, May 14, 1910–June 12, 1977,' National Academies Press Biographical Memoir, http://books.nap.edu/html/biomems/ilerner.html.
88 For information on Graham's education, his general work with poultry, and his breeding experiments, see Quirk, *Prof. William Richard Graham*; and Snyder, 'History of the Poultry Science Department at the Ontario Agricultural College.'
89 Hawthorne, 'A Five-Year Farm Management Survey,' 27, 29–31.
90 *American Poultry Journal*, December 1931, 5.
91 Select Committee, *Proceedings*, 519, 512.
92 USDA, 'Science in Farming,' *Yearbook*, 1943–7, 225.

Chapter 5: The 'Scientizing' of Breeding

1 'Hatchery Industry Basics – Past, Present & Future,' speech by Donald Shaver, 1978, Shaver Collection, series 4, box 19, file 38, Archival and Special Collections, University of Guelph Library.

2 R. Bradley, *The Country Gentleman and Farmer's Monthly Director* (London: D. Browne, 1732), 64–6.
3 O.A. Hanke, ed., *American Poultry History, 1823–1973* (Madison, WI: American Poultry History Society, 1974), 145; G. Sawyer, *The Agribusiness Poultry Industry: A History of Its Development* (New York: Exposition Press, 1971), 26.
4 *Farmer's Advocate*, 26 April 1923, 646.
5 Ontario Agricultural Commission, *Report*, 1880, appendix L, 6.
6 Hanke, ed., *American Poultry History*, 165; E.M. Funk, *Hatchery Operation and Management* (New York: John Wiley & Sons, Inc., 1955), 19–21; *American Poultry Journal*, May 1925, 570.
7 Hanke, ed., *American Poultry History*, 580–1.
8 *Farmer's Advocate*, 15 March 1928, 466 and 22 March 1928, 512; S. Leeson, *The Ontario Poultry Industry: An Illustrated History* (Ontario Poultry Council, 1986), 32, 45.
9 *Farm and Dairy*, 6 February 1913, 136.
10 See Lu Ann Jones, *Mama Learned Us to Work* (Chapel Hill: University of North Carolina Press, 2002), 86.
11 See *Farmer's Advocate*, 14 March 1912, 481; 2 May 1912, 841.
12 See *Farmer's Advocate*, West, 5 May 1903, 448; 14 April 1904, 531.
13 See *Farmer's Advocate*, 16 February 1905, 231.
14 Ibid., 20 April 1905, 585; 8 March 1906, 365.
15 Ibid., 4 March 1909, 349; 11 January 1912, 72.
16 *Farmer's Advocate*, West, 25 February 1922, 136.
17 *Farmer's Advocate*, 1 March 1923, 312, 313.
18 Ibid., 25 April 1922, 312.
19 Ibid., 4 February 1932, 69.
20 See *American Poultry Journal*, October 1911, 1429; March 1912, 618c.
21 *Canadian Poultry Review*, July 1907, 639.
22 *American Poultry Journal*, March 1910, 569.
23 Ibid., April 1925, 476–7, 502, 504–5, 506, 507, 509–12, 516–17.
24 Ibid., April 1925, 512.
25 Ibid., April, 1925, 510.
26 Ibid., July 1925, 719.
27 H.E. Botsford, *the Economics of Poultry Management* (New York: John Wiley & Sons, Inc., 1952), 120; *American Poultry Journal*, December 1931, 14.
28 Funk, *Hatchery Operation*, 21.
29 Sawyer, *The Agribusiness Poultry Industry*, 27.
30 *American Poultry Journal*, July 1921, 735.
31 *Farmer's Advocate*, 8 July 1937, 429; 14 July 1938, 427.
32 Ibid., 10 February 1938, 86.

33 Ibid., 15 March 1928, 466; 10 February 1938, 85.
34 Sawyer, *The Agribusiness Poultry Industry*, 26; *American Poultry Journal*, December 1910, 1446.
35 *American Poultry Journal*, February 1909, 149–51; March 1909, 293, 333.
36 Ibid., June 1922, 684.
37 Ibid., January 1911, 57–8.
38 See D. Fitzgerald, *Every Farm a Factory* (New Haven: Yale University Press, 2003), 24, 189.
39 See, for example, comments in *American Poultry Journal*, September 1926, 764–6, May 1927, 590, August 1927, 715, 717, 720, 721, and February 1932, 19; G.E. Bugos, 'Intellectual Property Protection in the American Chicken-breeding Industry,' *Business History Review* 66 (1992): 135.
40 *American Poultry Journal*, December 1910, 1450.
41 Ibid., May 1925, 602.
42 Ibid., November 1927, 866.
43 *Farmer's Advocate*, 15 March 1928, 466.
44 Ibid., 1 September 1927, 1275.
45 Ibid., 26 July 1928, 1169.
46 *American Poultry Journal*, July 1925, 716, 718.
47 *Farmer's Advocate*, 16 April 1925, 616.
48 *American Poultry Journal*, September 1925, 841–2.
49 Hanke, ed., *American Poultry History*, 170, 172.
50 *American Poultry Journal*, September 1925, 804.
51 Ibid., August 1927, 705. A long article on the British Columbia breeding of ROP birds appeared in the *American Poultry Journal* in April 1926, 435, 463–8. American poultrymen studied the BC situation carefully, and also looked into how the Canadian ROP system worked from Ottawa.
52 Ibid., August 1927, 715–16.
53 K.J. Cooke, 'Expertise, Book Farming, and Government Agriculture: The Origins of Agricultural Seed Certification in the United States,' *Agricultural History* 76 (2002): 524–45.
54 *American Poultry Journal*, September 1925, 804.
55 Ibid., September 1926, 737.
56 Ibid., September 1927, 750, 751.
57 Ibid., March 1930, 37.
58 Ibid., September, 1932, 9; Bugos, 'Intellectual Property Protection,' 135.
59 *American Poultry Journal*, September 1931, 11.
60 A.L. Hagedoorn and G. Sykes, *Poultry Breeding: Theory and Practice* (London: Crosby Lockwood & Son Ltd, 1953), 150–1, 217, 220, 222.
61 Ibid., 150–1, 220.

62 J. Marie, 'For Science, Love and Money: The Social Worlds of Poultry and Rabbit Breeding in Britain, 1900–1940,' *Social Studies of Science* 38 (2008): 926–7.
63 *American Poultry Journal*, June 1931, 5.
64 Hanke, ed., *American Poultry History*, 262–3, 702.
65 See L.A. Wilhelm, 'The Chick Buyer Is Boss,' Synopses of Addresses, Ontario Poultry Conference, 1953, box 1, Ontario Agricultural College, Dept of Poultry Husbandry, Archival and Special Collections, University of Guelph Library.
66 *Farmer's Advocate*, 5 July 1934, 401, 403; 27 September 1934, 539; and 28 March 1935, 165.
67 See, for example, ibid., 14 March 1935, 148, 20 June 1935, 349, 11 April 1935, 218, 21 November 1935, 659, 19 December 1935, 714, 715, 9 April 1936, 185, and 23 April 1936, 221, 239; and *American Poultry Journal*, March 1930, 64.
68 R.D. Crawford, ed., *Poultry Keeping and Genetics* (New York: Elsevier, 1990), 997, 1010–11.
69 *Farmer's Advocate*, 9 July 1936, 391, 25 March 1937, 190, 8 April 1937, 229, 22 April 1937, 263, 8 July 1937, 429, 25 November 1937, 701, 10 February 1938, 85; *American Poultry Journal*, May 1923, 655, March 1924, 456–62, November 1927, 867; *Breeder's Gazette*, 25 January 1923, 116.
70 *American Poultry Journal*, May 1923, 655; March 1924, 456–62; May 1925, 602.
71 *Farmer's Advocate*, 10 February 1938, 85.
72 Sawyer, *The Agribusiness Poultry Industry*, 112; E.L. Schapsmeier and F.H. Schapsmeier, *Henry A. Wallace of Iowa: The Agrarian Years, 1910–1940* (Ames: Iowa State University Press), 21, 27, 28.
73 See F.A. Hayes and G.T. Klein, *Poultry Breeding Applied* (Mount Morris, IL: Watt Publishing Co., 1953), 192–5; D.C. Warren, 'Techniques of Hybridization of Poultry,' *Poultry Science* 29 (1950): 60; Hagedoorn and Sykes, *Poultry Breeding*, 184–5.
74 P.B. Seigel et al., 'Genetic Selection Strategies – Population Genetics,' *Poultry Science* 76 (1997): 1064.
75 Interview with Donald McQueen Shaver, poultry breeder and founder of Shaver Poultry Breeding Farms, 23 April 2007. See, for example, *Advocate*, 11 January 1958, 37; 8 February 1958, 5; 14 February 1959, 3; 9 January 1960, 35.
76 Leeson, *The Ontario Poultry Industry*, 47.
77 See *Farmer's Advocate*, 11 March 1961, 44; D. Shaver, 'Hatchery Industry Basics – Past, Present, and Future,' Shaver Collection, series 4, box 19, file 38, Archival and Special Collections, University of Guelph.

78 E.S. Snyder, 'A History of the Poultry Science Department at the Ontario Agricultural College, 1894–1968' (unpublished manuscript, 1970), 289.
79 *Farmer's Advocate*, 26 June 1941, 426.
80 *Farmer's Advocate*, 23 January 1941, 50.
81 *Farmer's Advocate*, 11 January 1958, 37; 8 February 1958; 14 February 1958, 3; 9 January 1960, 35.
82 Hanke, ed., *American Poultry History*, 256.
83 *Farmer's Advocate*, 11 March 1961, 44.
84 USDA, Economic Research Service, 'Evolution of Vertical Coordination in the Poultry, Egg, and Pork Industries,' *Vertical Coordination of Marketing Systems / AER–807*, 2–3.
85 *Farmer's Advocate*, 14 July 1949, 574.
86 Hanke, ed., *American Poultry History*, 583; Leeson, *The Ontario Poultry Industry*, 27.
87 *Farmer's Advocate*, 28 June 1951, 23.
88 Shaver Poultry Breeding Farm, 1953, Shaver Collection, series 1, box 3, file 9, Archival and Special Collections, University of Guelph Library.
89 *Farmer's Advocate*, 28 June 1951, 23.
90 Hagedoorn and Skyes, *Poultry Breeding*, 219–22.
91 Ibid., 227, 228.
92 Ibid., 227–8.
93 J. Sapp, 'The Struggle for Authority in the Field of Heredity, 1900–1932: New Perspectives on the Rise of Genetics,' *Journal of the History of Biology* 16 (1983): 338.
94 F.B. Hutt and B.A. Rasmusen, 'Whither Poultry Genetics?' *Poultry Science* 21 (1965): 55.
95 See Y.V. Thaxton et al., 'The Decline of Academic Poultry Science in the United States of America,' *World's Poultry Science* 59 (2003): 303–13.
96 I.M. Lerner, *The Genetic Basis of Selection* (New York: John Wiley & Sons, Inc., 1958) 260.
97 See D. Fitzgerald, *The Business of Breeding: Hybrid Corn in Illinois, 1890–1940* (Ithaca: Cornell University Press, 1990).
98 Hutt and Rasmusen, 'Whither Poultry Genetics?' 60.
99 See T.B. Kinney Jr, 'Poultry Breeding Research in North America,' *World's Poultry Science Journal* 30–1 (1974–5): 8–27; and N.D. Bayley, 'Is There a Future for Land-Grant College Research and Extension?' *Poultry Science* 52 (1973): 5–15.
100 See Thaxton, 'The Decline of Academic Poultry Science.'
101 See S.L. Pardue, 'Education Opportunities and Challenges in Poultry Sci-

ence: Impact of Resource Allocation and Industry Needs,' *Poultry Science* 76 (1997): 938–43.
102 Snyder, 'History of the Poultry Science Department at the Ontario Agricultural College,' 232.
103 Agriculture Canada, *Reports*, 1949, 173; 1950, 192; 1952, 84; 1958, 94.
104 Agriculture Canada, *One Hundred Harvests: Research Branch Agriculture Canada, 1886–1986* (Ottawa: Agriculture Canada, 1986), 198.
105 Snyder, 'History of the Poultry Science Department at the Ontario Agricultural College, ' 289, 290, 291.
106 Interview of R.S. Gowe, director of Animal Research Centre, Research Branch, Agriculture Canada, 1965–86, with the author, 8 May 2007.
107 Ibid.
108 *Farmer's Advocate*, 11 March 1961, 44.

Chapter 6: The Rise of the Broiler Industry

1 O.A. Hanke, ed., *American Poultry History, 1823–1973* (Madison, WI: American Poultry History Society, 1974), 370–2.
2 Legislature of Ontario, Sessional Paper (henceforth Ontario, SP) 39, 1916, 39.
3 Ontario, SP 39, 1916, 42; *Farmer's Advocate*, 1 March 1901, 160, 15 April 1902, 305, 20 July 1916, 1227, and 29 August 1912, 1527; G. Sawyer, *The Agribusiness Poultry Industry: A History of Its Development* (New York: Exposition Press, 1971), 48.
4 Hanke, ed., *American Poultry History*, 363.
5 S. Leeson, *The Ontario Poultry Industry: An Illustrated History* (Ontario Poultry Council, 1986), 8, 11; Hanke, ed., *American Poultry History*, 419.
6 M.A. Jull et al., 'The Poultry Industry,' in United States Department of Agriculture, *Yearbook*, 1924, 402.
7 Ibid., 406.
8 *American Poultry Journal*, October 1922, 917, 951–2.
9 Ibid., November 1927, 864, 866.
10 See, for example, *Farming World*, 1 March 1907, 226; and *Farm and Dairy*, 1 February 1917, 104.
11 Canada Census, 1901, vol. 2, 252–3.
12 *Farming World and Canadian Farm and Home*, 2 March 1903, 83.
13 J.H. Hare and T.A. Benson, 'Farm Poultry and Egg Marketing Conditions in Ontario County,' Bulletin 208, Ontario Department of Agriculture, 1913, 6.

14 Canada Census, vol. 4, 1921, 66–7.
15 B. Theunissen, 'Breeding without Mendelism: Theory and Practice of Dairy Cattle Breeding in the Netherlands, 1900–1950,' *Journal of the History of Biology* 41 (2008): 641.
16 For the relationship of purebred beef cattle to purebred dairy cattle in North America, see M.E. Derry, *Ontario's Cattle Kingdom: Purebred Breeders and Their World, 1870–1920* (Toronto: University of Toronto Press, 2001), 93–5.
17 M.E. Derry, *Horses in Society: A Story of Breeding and Marketing Culture, 1800–1920* (Toronto: University of Toronto Press, 2006).
18 R.J. Wood and V. Orel, *Genetic Prehistory in Selective Breeding: A Prelude to Mendel* (Oxford: Oxford University Press, 2001); M.L. Ryder, *Sheep and Man* (London: Duckworth, 1983) and his 'The History of Sheep Breeds in Britain,' *Agricultural History Review* 12 (1964): part 1: 1–12, part 2: 79–97.
19 V. McCormick, 'Butter and Egg Business: Implications from the Records of a Nineteenth-century Farm Wife,' *Ohio History* 100 (1991): 329–35, and her *Farm Wife: A Self-Portrait, 1886–1896* (Ames: Iowa State University Press, 1990).
20 See S. McMurry, *Transforming Rural Life: Dairying Families and Agricultural Change, 1820–1885* (Baltimore: Johns Hopkins University Press, 1995).
21 There is a great deal that has been written on dairying and women. See McMurry, *Transforming Rural Life*. For the issue of women and, in particular, butter, the last stand of independence for women in dairying in North America, see M.E. Derry, 'Gender Conflicts in Dairying: Ontario's Butter Industry, 1880–1920,' *Ontario History* 90 (1998): 31–47.
22 Derry, *Ontario's Cattle Kingdom*, 93–5, 107–15.
23 *Farmer's Advocate*, 5 September 1912, 1558–9.
24 See, for example, *Farmer's Advocate*, 14 March 1912, 481; 11 April 1912, 688–9.
25 *Farmers' Advocate*, 14 December 1916, 2046.
26 See, for example, Lu Ann Jones, *Mama Learned Us to Work* (Chapel Hill: University of North Carolina Press, 2002), 94; and McCormick, 'Butter and Egg Business,' 63–4.
27 Jones, *Mama Learned Us to Work*, 80, 85, 87.
28 See Canada Census, 1901, vol. 2, 203; Canada Census, 1911, vol. 4, 410, 418; Canada Census, vol. 4, 1921, ci, 61, 66–7.
29 See Derry, *Ontario's Cattle Kingdom* and *Horses in Society* for an overall assessment of Canadian and American views of livestock experts on the subject. Numerous complaints about farmers' lack of interest in pure breeding can be found in the *Farmer's Advocate* and the *Breeder's Gazette*.
30 *Canadian Poultry Review*, March 1892, 39.

31 Ibid., June 1892, 91.
32 *American Poultry Journal*, October 1922, 917, 951–2.
33 Ibid., 951.
34 In Canada, for example, Leghorns outnumbered Rocks by 1921. Canada Census, vol. 4, 1921, 66–7.
35 Sawyer, *The Agribusiness Poultry Industry*, 37.
36 *Farmer's Advocate*, December 1868, 185.
37 Sawyer, *The Agribusiness Poultry Industry*, 113–14.
38 *American Poultry Journal*, December 1931, 21.
39 Hanke, ed., *American Poultry History*, 403–8.
40 See H.L. Shrader, 'The Chicken-of-Tomorrow Program: Its Influence on "Meat-Type" Poultry Production,' *Poultry Science* 31 (1952) 3–8; R. Horowitz, 'Making the Chicken of Tomorrow: Reworking Poultry as Commodities and as Creatures, 1945–1990,' in *Industrializing Organisms: Introducing Evolutionary History*, ed. S.R. Schrepfer and P. Scranton (London: Routledge, 2004), 217–21; Hanke, ed., *American Poultry History*, 400–7 ; G.E. Bugos, 'Intellectual Property Protection in the American Chicken-breeding Industry,' *Business History Review* 66 (1992): 127–68; J. Biely, H.C. Gasperdone, and W.H. Pope, 'Broiler Production: 25 Years of Progress (Canada versus U.S.A.),' *World's Poultry Science Journal* 27 (1971): 241–61.
41 Shrader, 'The Chicken-of-Tomorrow Program,' 7–8.
42 Leeson, *The Ontario Poultry Industry*, 71–2, 74.
43 See, for example, *Farmer's Advocate*, 26 October 1933, 600–1; 21 November 1935, 659; 10 March 1938, 121; 9 March 1939, 121; 13 July 1939, 409; 25 August 1925, n.p.; and 22 August 1959, n.p.
44 Leeson, *The Ontario Poultry Industry*, 71–2, 74.
45 Interview with Donald McQueen Shaver, poultry breeder and founder of Shaver Poultry Breeding Farms, 23 April 23 2007.
46 US Department of Agriculture, 'Better Plants and Animals,' part 1, *Yearbook of Agriculture*, 1936; USDA, Economic Research Service, 'Evolution of Vertical Coordination in the Poultry, Egg, and Pork Industries,' *Vertical Coordination of Marketing Systems / AER–807*, 2–3.
47 M.R. Gisolfi, 'From Crop Lien to Contract Farming: The Roots of Agribusiness in the American South, 1929–1939,' *Agricultural History* 80 (2006): 167, 168, 176, 177–8, 181.
48 Sawyer, *The Agribusiness Poultry Industry*, 87, 90, 91.
49 Ibid., 133.
50 S. Strittler, *Chicken: The Dangerous Transformation of America's Favorite Food* (New Haven: Yale University Press, 2005), 45.

51 Leeson, *The Ontario Poultry Industry*, 23, 88–102; *Farmer's Advocate*, 26 August 1943, 514, 26 July 1945, 540, 14 July 1949, 574, 10 August 1950, 619, and 22 February 1951, 36.
52 Urquhart and Buckley, *Historical Statistics of Canada*, 367.
53 Leeson, *The Ontario Poultry Industry*, 37.
54 *Farmer's Advocate*, 23 February 1959, 39.
55 Striffler, *Chicken: the Dangerous Transformation*, 47, 48; A.W. Jasper, 'The Farmer and the Poultry Industry,' *World's Poultry Science Journal* 27 (1971): 43–9.
56 Gisolfi, 'From Crop Lien to Contract Farming,' 181.
57 Sawyer, *The Agribusiness Poultry Industry*, 83, 84, 133, 137; Hanke, ed., *American Poultry History*, 418–20.
58 R.D. Crawford, ed., *Poultry Keeping and Genetics* (New York: Elsevier, 1990), 985.
59 Urquhart and Buckley, *Historical Statistics of Canada*, 380.
60 Sawyer, *The Agribusiness Poultry Industry*, 215.
61 C.K. Laurent, 'The Effect of Integration in the Broiler Industry,' Synopses of addresses, Ontario Poultry Conference, 1957, 53, 54, 56, box 1, OAC, Department of Poultry Husbandry, Archives, University of Guelph.
62 *Farmer's Advocate*, 14 February 1959, 3.
63 *Farmer's Advocate*, 28 November 1964, 16.
64 I.K. Felch, *The Breeding and Management of Poultry* (Hyde Park, NY: Norfolk County Press, 1877), 20, 22.
65 H.W. Hawthorne, 'A Five-year Farm Management Survey in Palmer Township, Washington County, Ohio, 1912–1916,' Bulletin 716, US Department of Agriculture, September 1916, 27, 29–31.
66 Jones, *Mama Learned Us to Work*, 85.
67 Hanke, ed., *American Poultry History*, 218.
68 *Farming World and Canadian Farm and Home*, 12 March 1901, 677.
69 Ibid., 2 March 1903, 77.
70 *Proceedings* of the Select Committee into Agricultural Conditions, vol. 1, 1924, Ottawa 519, 512.
71 *Farmer's Advocate*, 20 February 1930, 270.
72 Ibid., 26 January 1939, 32.
73 Ibid., 12 March 1955, 50.
74 Hanke, ed., *American Poultry Industry*, 218, 240, 245.
75 Gisolfi, 'From Crop Lien to Contract Farming,' 178, 179, 180.
76 F.D. Hutt, 'Seventy-five Years of Poultry Genetics,' Roundtable of Poultry Breeders Association, 1975, at www.poultryscience.org/pba/1952-2003/1975/1975%20Hutt.pdf.
77 See his (along with E.R. Dempster) 'Some Aspects of Evolutionary Theory

in the Light of Recent Work on Animal Breeding,' *Evolution* 2 (1948): 19–28, and (along with L.N. Hazel) 'Population Genetics of a Poultry Flock under Artificial Selection,' *Genetics* 32 (1947): 325–39.
78 R.W. Allard, 'Israel Michael Lerner, May 14, 1910–June 12, 1977,' National Academies Press Biographical Memoir, National Academy of Sciences, http://books.nap.edu/html/biomems/ilerner.html Accessed 18 February 2007.
79 I.M. Lerner and H. Donald, *Modern Developments in Animal Breeding* (London and New York: Academic Press, 1966), 170, 171, 178–9, 185.
80 See Theunissen, 'Breeding without Mendelism,' 659, 660, 664, 668.
81 A.L. Hagedoorn and G. Skyes, *Poultry Breeding: Theory and Practice* (London: Crosby Lockwood & Son Ltd, 1953), 14
82 A. Fraser, *Animal Breeding Heresies* (London: Crosby Lockwood & Son Ltd, 1960), 7–8, 10, 11.

Chapter 7: Epilogue

1 R.B. Talbot, *The Chicken War: An International Trade Conflict between the United States and the European Community, 1961–64* (Ames: Iowa State University Press, 1978), 3–19.
2 S. Striffler, *Chicken: The Dangerous Transformation of America's Favorite Food* (New Haven: Yale University Press, 2005), 17–19.
3 Ibid., 19.
4 Ibid., 17–19.
5 A short, interesting review of genetics and poultry breeding appears in G. Sawyer, *The Agribusiness Poultry Industry: A History of Its Development* (New York: Exposition Press, 1971), 111–24.
6 G.E. Bugos, 'Intellectual Property Protection in the American Chicken-breeding Industry,' *Business History Review* 66 (1992): 127–68.
7 'Primary Poultry Breeders: Company Profiles,' at http://www.agr.gc.ca/poultry/brpr-elpr_m.htm.
8 K. Laughlin, 'The Evolution of Genetics, Breeding, and Production,' Temperton Fellowship Report no. 15, Harper Adams University College, UK, June 2007, 35, 18, 36.
9 For information on B.C. Partners see 'B.C. Partners – About Us,' at http://www.bcpartners.com/bcp/aboutus; and 'B.C. Partners – Case Studies,' at http://bcpartners.com/bcp/cases/.
10 'Aviagen,' at http://www.aviagen.com/output.aspx?sec=2040&con=982&siteld=1; 'Advent International: News Article,' at http://adventinternational.fr/News/Article.aspx?PageID=7.1&Ne; interview

with Dominic Elfick, animal geneticist, and director of research and development, Aviagen, 9 May 2007; 'Aviagen Becomes Part of the Erich Wesjohann Group,' at http://www.aviagen.com/output.aspx?sec=338&con=3056.

11 See J.L. Lush, 'Improving One Character by Breeding for Another,' Fact Finding Conference of the Institute of American Poultry Industries, 1958–9, 184–5; T.B. Kinney Jr, 'Poultry Breeding Research in North America,' *World's Poultry Science Journal* 30–1 (1974–5): 8.

12 See Bugos, 'Intellectual Property Protection in the American Chicken-breeding Industry'; F.D. Hutt, 'Seventy-five Years of Poultry Genetics,' Roundtable of Poultry Breeders Association, 1975, at www.poultryscience.org/pba/1952-2003/1975/1975%20Hutt.pdf.

13 A. Heisdorf, 'Twenty Years Experience with Reciprocal Recurrent Selection,' Eighteenth Annual Session, National Poultry Breeders' Roundtable, 1969, 115–16.

14 See M.A. Jull, *Poultry Breeding*, 3rd ed. (New York: John Wiley & Sons, Inc., 1952), 375–6; A.E. Bell et al., 'Systems of Breeding Designed to Utilize Heterosis in Domestic Fowl,' *Poultry Science* 31 (1952): 11–13; C.F. McClary, 'Reciprocal Recurrent Selection Response in Poultry, Other Animals and Plants,' *Eighteenth Annual Session*, National Poultry Breeders' Roundtable, 1969, 120–55; Heisdorf, 'Twenty Years Experience,' 114–19; and F.N. Jerome, 'A Discussion of the Breeding Plan – Recurrent Reciprocal Selection,' 1969, series 1, box 2, file 14, Shaver Collection, Archival and Special Collections, University of Guelph Library; F.A. Hays and G.T. Klein, *Poultry Breeding Applied* (Mount Morris, IL: Watt Publishing Co., 1952), 196–8.

15 Hayes and Klein, *Poultry Breeding Applied*, 197–8; Jerome, 'A Discussion of the Breeding Plan.'

16 O.A. Hanke, ed., *American Poultry History, 1823–1973* (Madison, WI: American Poultry History Society, 1974), 271.

17 C.F. McClary, 'Opening Remarks,' Ninth Annual Session, National Poultry Breeders' Roundtable, 1960, 2; A.W. Nordskog, 'The Evolution of Animal Breeding Practices – Commercial and Experimental,' Fourteen Annual Session, National Poultry Breeders' Roundtable, 1965, 58–9, 72.

18 For good assessments of the fundamentals about the way breeding proceeded by 1990 see R.D. Crawford, ed., *Poultry Keeping and Genetics* (New York: Elsevier, 1990), 987–1009; and L. Stevens, *Genetics and Evolution of the Domestic Fowl* (Cambridge: Cambridge University Press, 1991), 146–63.

19 G.B. Havenstein, 'Performance Changes in Livestock Following 50 Years of Genetic Selection,' *Lohmann Information* 41 (December 2006): 30.

20 See, for example, J. Biely, H.C. Gasperdone, and W.H. Pope. 'Broiler

Production: 25 Years of Progress (Canada versus U.S.A.),' *World's Poultry Science Journal* 27 (1971): 241–61.
21 Kinney, 'Poultry Breeding Research in North America,' 8–10.
22 Interview with Robert S. Gowe, animal geneticist, former director of the Animal Research Centre, Research Branch, Agriculture Canada, and subsequently a geneticist for Shaver Poultry Breeding Farm, 8 May 2007; interview with John Hardiman, animal geneticist, and vice-president of research and development, Cobb-Vantress, Inc., 8 May 2007; W.A. Rishell, 'Breeding and Genetics. Symposium: Genetic Selection – Strategies for the Future, Breeding and Genetics – Historical Perspective,' *Poultry Science* 76 (1997): 1057–61; G. Bulfield, 'Strategies for the Future,' *Poultry Science* 76 (1997): 1071; Laughlin, 'The Evolution of Genetics, Breeding, and Production,' 27.
23 Laughlin, 'The Evolution of Genetics, Breeding, and Production,' 26, 49.
24 Interesting studies have recently been done in how well AI can work with quantitative genetics. See S. Wilmot, 'From "Public Service" to Artificial Insemination: Animal Breeding Science and Reproductive Research in Early 20th Century Britain,' *Studies in History and Philosophy of Biological and Biomedical Sciences* 38 (2007): 411–41, and her 'Between the Farm and the Clinic: Agricultural and Reproductive Technology in the Twentieth Century,' ibid., 303–15. See also T. Theunissen, 'Breeding without Mendelism: Theory and Practice of Dairy Cattle Breeding in the Netherlands, 1900–1950,' *Journal of the History of Biology* 41 (2008): 637–76, and his 'The Holsteinization of the Friesians: Culture of Dairy Cattle Breeding in the Netherlands, 1945–1995,' *Isis*, forthcoming; C. Grasseni, 'Managing Cows: An Ethnography of Breeding Practices and Uses of Reproductive Technology in Contemporary Dairy Farming in Lombardy (Italy),' *Studies in History and Philosophy of Biological and Biomedical Sciences* 38 (2007): 488–510.
25 Havenstein, 'Performance Changes in Livestock,' 30, 35.
26 G.B. Havenstein et al., 'Growth, Livability, and Feed Conversion of 1957 versus 2001 Broilers when Fed Representative 1957 and 2001 Broiler Diets,' *Poultry Science* 82 (2003): 1500–8.
27 'Poultry Marketplace – Overview of the Primary Poultry Breeder Industry,' at http://www.agr.ca/poultry/brov-elap_e.htm; 'Aviagen,' at http://www.aviagen.com/output.aspx?sec=2040&con=982&siteld=; 'The Impact of Genetics on Breeder Management,' May 2005 article in *Hybro B.V.*, reproduced on *The Poultry Site*, http://www.thepoultrysite.com/FeaturedArticle/FAType.asp?AREA.
28 Interviews with John Hardiman; Douglas W. Dodds, former chief executive officer of Schneider Foods, and currently strategic consultant, Maple Leaf Foods, 10 April 2007; Dominic Elfick; Ken Laughlin, wildlife biologist

and geneticist, group vice-president, policy and strategy, Aviagen, 24–5 May 2007. Shaver Collection, series 1, box 3, file 1, Archival and Special Collections, University of Guelph.
29 Interview with Dominic Elfick.
30 See, for example, Bugos, 'Intellectual Property Protection in the American Chicken-Breeding Industry.'
31 See, for example, Shaver Collection, series 1, box 3, file 1, Archival and Special Collections, University of Guelph.
32 Interview with Dominic Elfick.
33 Crawford, ed., *Poultry Keeping and Genetics*, 1029, 1030–1, 1037–8, 1041, 1042–3, 1047.
34 Ibid., 1029.
35 Ibid., 1047–8.
36 *Poultry Tribune*, August 1985, Shaver Collection, series 1, box 1, file 1, Archival and Special Collections, University of Guelph.
37 Shaver Collection, series 1, box 1, file 1, Archival and Special Collections, University of Guelph.
38 Interview with Donald McQueen Shaver, poultry breeder and founder of Shaver Poultry Breeding Farms, 23 April 2007.
39 *Shaver News*, August 1985, Shaver Collection, series 1, box 1, file 1, Archival and Special Collections, University of Guelph.
40 Shaver Collection, series 1, box 3, file 9, Archival and Special Collections, University of Guelph.
41 Interview with Shaver.
42 *Poultry World*, 15 October 1970, Shaver Collection, series 1, box 3, file 9, Archival and Special Collections, University of Guelph.
43 Shaver Collection, series 1, box 4, file 7, Archival and Special Collections, University of Guelph.
44 'Breeding Plans,' 1959, Shaver Collection, series 1, box 2, file 1, Archival and Special Collections, University of Guelph.
45 Cole to Shaver, 7 April 1968, Shaver Collection, series 1 box 4, file 7, Archival and Special Collections, University of Guelph.
46 *Poultry Tribune*, August 1985.
47 Gowe to Shaver, series 1, box 1, file 2, Shaver Collection, Archival and Special Collections, University of Guelph.
48 For more on breeding, vaccines, and disease resistance, see Crawford, ed., *Poultry Keeping and Genetics*, 996.
49 Cole to Shaver, 6 December 1974, Shaver Collection, series 1, box 4, file 5, Archival and Special Collections, University of Guelph.
50 Series 1, box 2, file 1, Shaver Collection, Archival and Special Collections, University of Guelph.

51 D.K. Flock and R. Preisinger, 'Specialization and Concentration as Contributing Factors to the Success of the Poultry Industry in the Global Food Market,' Draft for *Lohmann Information*, winter 2007, 4.
52 *Poultry Tribune*, August 1985.
53 Series 1, box 3, file 9, Shaver Collection, Archival and Special Collections, University of Guelph.
54 Interviews with Shaver and Robert S. Gowe.
55 Shaver to Cole, 8 April 1968, series 1, box 4, file 7, Shaver Collection, Archival and Special Collections, University of Guelph.
56 *Broiler Business*, January 1977, series 1, box 2, file 1, Shaver Collection, Archival and Special Collections, University of Guelph.
57 *Shaver News*, August 1985.
58 Interview with James Wilton, animal geneticist, University of Guelph, 18 April 2007; Hanke, ed., *American Poultry History*, 256.
59 Interview with Shaver.
60 Series 1, box 11, file 4, Shaver Collection, Archival and Special Collections, University of Guelph.
61 See US Department of Agriculture, 'Better Plants and Animals,' part 1, *Yearbook* of Agriculture, 1936, 849.
62 Series 1, box 11, file 5, Shaver Collection, Archival and Special Collections, University of Guelph.
63 New Rules Project, *The Agriculture Sector*, 'Canadian Supply Management System,' http://www.newrules.org/agri/CanadaSM.html; 'When the Farmer Makes the Rules,' http://www.newrules.org/journal/nrfall100farmer.html; 'Supply Management,' http://www.newrules.org/agri/splymg2.html; 'US Supply Management Proposals,' http://www.newrules.org/agri/USsplmgmt.html.
64 S. Leeson, *The Ontario Poultry Industry: An Illustrated History* (Ontario Poultry Council, 1986), 105–7.
65 Ibid., 107, 113.
66 New Rules Project, *The Agriculture Sector*, 'Canadian Supply Management System'; 'When the Farmer Makes the Rules'; 'Supply Management.'
67 Interviews with Douglas W. Dodds, 10 April and 3 May 2007.
68 Crawford, ed., *Poultry Keeping and Genetics*, 897, 899, 901, 909.
69 Interview with Dominic Elfick.
70 Chicken Farmers of Canada, http://www.chicken.ca; Chicken Farmers of Canada, 'Animal Care Fact Sheet,' http://www.chicken.ca/DefaultSite/index_e.aspx?DetailID=16; Chicken Farmers of Ontario, *Supply Management: A Recipe for Success* (Burlington, ON: 2003).
71 'Shane Report: Canadian Industry at Crossroads,' *Watt Poultry.Com*, http://www.wattpoultry.com/Article.aspx?id=12148.

72 New Rules Project, 'When the Farmer Makes the Rules.'

Conclusion

1 See V.B. Smocovitis, *Unifying Biology: The Evolutionary Synthesis and Evolutionary Biology* (Princeton: Princeton University Press, 1996); and A. Serafini, *The Epic History of Biology* (New York: Plenum Press, 1993).
2 See, for example, W. Provine, *Sewell Wright and Evolutionary Biology* (Chicago: University of Chicago Press, 1986), 36–7, 43, 54, 135, 141, 159. See also W.E. Castle, *Heredity in Relation to Evolution and Animal Breeding* (New York: D. Appleton and Co., 1911), 151.
3 I.K. Felch, *Poultry Culture: How to Raise, Mate and Judge Thoroughbred Fowls* (Chicago: Donohue, Henneberry, 1902), 45. Compare Felch with D.C. Warren, *Practical Poultry Breeding* (New York: Macmillan Co., 1953); and A. Jull, *Poultry Breeding*, 3rd ed. (New York: John Wiley & Sons, Inc., 1952).
4 R.D. Crawford, ed., *Poultry Keeping and Genetics* (New York: Elsevier, 1990), 958.
5 Ibid., 973.
6 See T.B. Kinney, Jr, 'Poultry Breeding Research in North America,' *World's Poultry Science Journal* 30–1 (1974–5): 9–27.
7 S. Wright, *Evolution and the Genetics of Populations*, vol. 3, *Experimental Results and Evolutionary Deductions* (Chicago: University of Chicago Press, 1977), 30.
8 Crawford, ed., *Poultry Keeping and Genetics*, 913, 914.
9 O. Harman and M.R. Dietrich, eds, *Rebels, Mavericks, and Heretics in Biology* (New Haven: Yale University Press, 2008), 265–78.
10 Crawford, ed., *Poultry Keeping and Genetics*, 913, 914.
11 F.D. Hutt and B.A. Rasmusen, *Animal Genetics*, 2nd ed. (New York: John Wiley & Sons, 1982), 393.
12 See N.M. Springer and R.M. Stupar, 'Allelic Variation and Heterosis in Maize: How Do Two Halves Make More than a Whole?' *Genome Research* 17 (2007): 264–75; A.F. Troyer, 'Adaptedness and Heterosis in Corn and Mule Hybrids,' *Crop Science* 46 (2006): 528–43; A.R. Hallauer, 'History, Contribution, and Future of Quantitative Genetics in Plant Breeding: Lessons from Maize,' *Crop Science* 47 (2007), from *International Plant Breeding Symposium*, 2007, S4–19.
13 J.A. Birchler et al., 'Unraveling the Genetic Basis of Hybrid Vigor,' *Proceedings of the National Academy of Sciences of the USA* 103 (2006): 12957. See as well J.A. Birchler et al., 'In Search of the Molecular Basis of Heterosis,' *The Plant Cell* 15 (2003): 2236–9.

14 D.B. Paul and B.A. Kimmelman, 'Mendel in America: Theory and Practice, 1900–1919,' in *The Development of American Biology*, ed. R. Rainger et al. (Philadelphia: University of Pennsylvania Press, 1988), 296, 299.
15 See S. Wilmot, 'From "Public Service" to Artificial Insemination: Animal Breeding Science and Reproductive Research in Early 20th Century Britain,' *Studies in History and Philosophy of Biological and Biomedical Sciences* 38 (2007): 411–41, and her 'Between the Farm and the Clinic: Agricultural and Reproductive Technology in the Twentieth Century,' ibid., 303–15. T. Theunissen, 'Breeding without Mendelism: Theory and Practice of Dairy Cattle Breeding in the Netherlands, 1900–1950,' *Journal of the History of Biology* 41 (2008): 637–76, and his 'The Holsteinization of the Friesians: Culture of Dairy Cattle Breeding in the Netherlands, 1945–1995,' *Isis*, forthcoming paper. C. Grasseni, 'Managing Cows: An Ethnography of Breeding Practices and Uses of Reproductive Technology in Contemporary Dairy Farming in Lombardy (Italy),' *Studies in History and Philosophy of Biological and Biomedical Sciences* 38 (2007): 488–510.

Bibliography

Primary Sources

MANUSCRIPT COLLECTIONS

Donald McQueen Shaver Collection, series 1 (business records), 3 (personal papers), and 4 (addresses, speeches, and presentations), Archives, University of Guelph.
Ontario Agricultural College Department of Poultry Husbandry, Archives, University of Guelph.

GOVERNMENT DOCUMENTS

Agriculture Canada. *One Hundred Harvests: Research Branch Agriculture Canada, 1886–1986*. Ottawa, 1986.
– *Reports*.
Bergey, J.E. 'Getting Rid of the Loafer Hen.' Circular 55, Manitoba Department of Agriculture, July 1919.
Board of Inquiry into the Cost of Living. *Report*. 2 vols. Ottawa, Canada, 1915.
Canada. Parliament. Sessional Papers.
Foley, A.W. 'Practical Poultry Keeping.' Poultry Bulletin 2, Alberta Department of Agriculture, 1911.
Funk, W.C. 'Value to Farm Families of Food, Fuel, and Use of House.' Bulletin 410, US Department of Agriculture, November 1916.
Graham, W.R. 'Farm Poultry.' Bulletin 127, Ontario Agricultural College, 1903.
Great Britain. Ministry of Agriculture, Fisheries, and Food. *General View of the Agriculture of Northumberland, Cumberland and Westmoreland, by J. Bailey and G. Culley*. Newcastle on Tyne: F. Graham, 1972.

Hare, J.H., and T.A. Benson. 'Farm Poultry and Egg Marketing Conditions in Ontario County.' Bulletin 208, Ontario Department of Agriculture, 1913.

Hawthorne, H.W. 'A Five-year Farm Management Survey in Palmer Township, Washington County, Ohio, 1912–1916.' Bulletin 716, US Department of Agriculture, September 1916.

Jull, M.A., et al. 'The Poultry Industry.' In US Department of Agriculture, *Yearbook*, 1924: 377–456.

McGrew, T.F. 'American Breeds of Poultry.' In US Department of Agriculture, Bureau of Animal Industry, *Report*, 1901: 513–65.

Ontario Agricultural Commission. *Report*. 1880.

Ontario Legislature. Sessional Papers.

Pearl, R. 'Breeding Poultry for Egg Production.' Bulletin 192, in *Annual Report of the Maine Agricultural Experiment Station*, 1911.

Richey, F.D. 'Corn Breeding.' Bulletin 1489, US Department of Agriculture, November 1927.

Select Committee on the Cost of Living. *Proceedings*. Ottawa, Canada, 1919.

Select Committee of the House of Commons into Agricultural Conditions. *Proceedings*. Ottawa, Canada, 1924.

Slocum, R.R. 'The Value of the Poultry Show.' In *Report*, Bureau of Animal Industry, US Department of Agriculture, 1908: 357–63.

United Kingdom. *Third Report of the Royal Commission on Horse Breeding*. 1890.

United States Department of Agriculture. *Reports*.
- *Yearbook*, 1936. 'Better Plants and Animals,' book 1.
- *Yearbook*, 1937. 'Better Plants and Animals,' book 2.
- *Yearbook*, 1943–7 [one vol.]. 'Science in Farming.'

United States Department of Agriculture. Economic Research Service. 'Evolution of Vertical Coordination in the Poultry, Egg, and Pork Industries.' *Vertical Coordination of Marketing Systems / AER-807*. 2002.

Warren, D.C. 'The Progeny Test in Poultry Breeding.' *Circular 168*, Kansas Agricultural Experiment Station, August 1932.

Wright, S. 'The Effects of Inbreeding and Crossbreeding on Guinea Pigs.' Bulletin 1090, US Department of Agriculture, November 1922.

JOURNALS AND NEWSPAPERS

Agricultural Gazette of Canada
American Poultry Journal
Breeder's Gazette
British Poultry Science
Canada Poultry Journal

Canadian Poultry Chronicle
Canadian Poultry Review
Farm and Dairy & Rural Home. 1908–18. Before 1908: *Farming World and Canadian Farm and Home*, 1903–7; *Farming World, for Farmers and Stockmen*, 1900–2; *Farming*, 1895–1900; *Canadian Live-Stock and Farm Journal*, 1886–95
Farmer's Advocate
O.A.C. Review (journal of the Ontario Agricultural College). 1880–1920
Poultry Science
World's Poultry Science Journal

INTERVIEWS

Dodds, Douglas W. Former chief executive officer of Schneider Foods, and currently strategic consultant, Maple Leaf Foods. 10 April, 3 May 2007.
Elfick, Dominic. Animal geneticist, and director of research and development, Aviagen. 9 May 2007.
Gowe, Robert S. Animal geneticist, former director of the Animal Research Centre, Research Branch, Agriculture Canada, and subsequently a geneticist for Shaver Poultry Breeding Farm. 8 May 2007.
Hardiman, John. Animal geneticist, and vice-president of research and development, Cobb-Vantress, Inc. 8 May 2007.
Laughlin, Ken. Wildlife biologist and geneticist, group vice-president, policy and strategy, Aviagen. 24–5 May 2007.
Shaver, Donald McQueen. Poultry breeder and founder of Shaver Poultry Breeding Farms. 23 April 2007.
Vanclief, Lyle. Minister of agriculture, Government of Canada, 1997–2003. 15 May 2007.
Wilton, James. Animal geneticist, University of Guelph. 18 April 2007.

MONOGRAPHS AND ARTICLES

Babcock, E.B., and R.E. Clausen. *Genetics in Relation to Agriculture.* New York: McGraw-Hill Book Co., Inc., 1918.
Bailey, J., and G. Culley. *General View of the Agriculture of Northumberland, Cumberland and Westmoreland.* London: B. McMillan, 1805; repr. Newcastle upon Tyne: Frank Graham, 1972.
Bajema, C.J., ed. *Artificial Selection and the Development of Evolutionary Theory.* Benchmark Papers in Systematic and Evolutionary Biology 4. Stroudsburg, PA: Hutchinson Ross Publishing Co., 1982.
Bakewell Letters – Culley and British Museum Collections. Part 2 in H.C. Paw-

son, *Robert Bakewell: Pioneer Livestock Breeder*. London: Crosby Lockwood & Son, Ltd, 1957.
Bayley, N.D. 'Is There a Future for Land-Grant College Research and Extension?' *Poultry Science* 52 (1973): 5–15.
Bell, A.E. 'Heritability in Retrospect.' *Journal of Heredity* 68 (1977): 297–300.
Bell, A.E., et al. 'Systems of Breeding Designed to Utilize Heterosis in the Domestic Fowl.' *Poultry Science* 31 (1952): 11–22.
Biely, J., H.C. Gasperdone, and W.H. Pope. 'Broiler Production: 25 Years of Progress (Canada versus U.S.A.).' *World's Poultry Science Journal* 27 (1971): 241–61.
Birchler, J.A., et al. 'In Search of the Molecular Basis of Heterosis.' *The Plant Cell* 15 (2003): 2236–9.
– 'Unraveling the Genetic Basis of Hybrid Vigor.' *Proceedings of the National Academy of Sciences of the USA* 103 (2006): 12957–8.
Blakeslee, A.F. 'Fancy Points vs. Utility.' *Journal of Heredity* 6 (1915): 175–81.
Botsford, H.E. *The Economics of Poultry Management*. New York: John Wiley & Sons, Inc., 1952.
Bradley, R. *The Country Gentleman and Farmer's Monthly Director*. London: D. Browne, 1732.
Brewbaker, J.L. *Agricultural Genetics*. Englewood Cliffs, NJ: Prentice-Hall, Inc., 1964.
Brown, E. *Poultry Breeding and Production*. Vol. 1. London: Ernest Benn Limited, 1929.
– *Races of Domestic Poultry*. London: Edward Arnold, 1906.
Brown, W.A. 'The Poultry Industry in Maine.' *O.A.C. Review* 22 (1910): 239–45.
Bulfield, G. 'Strategies for the Future.' *Poultry Science* 76 (1997): 1071–4.
Burnham, G. *Burnham's New Poultry Book*. Boston: Lee & Shepard, 1877.
– *The History of the Hen Fever*. San Diego: Frank E. Marcy, 1935; repr. of 1855 edition.
Burt, D.W. 'Chicken Genome: Current Status and Future Opportunities.' *Genome Research* 15 (2005): 1692–8.
Castle, W.E. *Heredity in Relation to Evolution and Animal Breeding*. New York: D. Appleton and Co., 1911.
– 'Pure Lines and Selection.' *Journal of Heredity* 5 (1914): 93–7.
– 'Some Biological Principles of Animal Breeding.' *American Breeders' Magazine* 3 (1912): 270–82.
Chicken Farmers of Canada. http://www.chicken.ca.
Chicken Farmers of Ontario. *Supply Management: A Recipe for Success*. Burlington, ON, 2003.
Commission on Genetic Resources for Food and Agriculture. 'The State of the

World's Animal Genetic Resources for Food and Agriculture.' Report, June 2007.

Crawford, R.D., ed. *Poultry Breeding and Genetics*. New York: Elsevier, 1990.

Culley, G. *Observations on Live Stock: Containing hints for choosing and improving the best breeds of the most useful kinds of domestic animals*. London: D. Longworth, 1804.

Darrah, L.B. *Business Aspects of Commercial Poultry Farming*. New York: Ronald Press Co., 1952.

Darwin, C. *On the Origin of Species*. London: John Murray, 1859.

– *The Variation of Animals and Plants under Domestication*. Foreword by Harriet Ritvo. Baltimore: Johns Hopkins University Press, 1998 (repr. of 1883 edition, first published 1868).

Davenport, C.B. 'The Relationship of the Association to Pure Research.' *American Breeders' Magazine* 1 (1911): 66–7.

Dempster, E.R., and I.M. Lerner. 'Heritability of Threshold Characters.' *Genetics* 35 (1950): 212–37.

East, E.M. 'Heterosis.' *Genetics* 21 (1936) 375–97.

Emmerson, D.A. 'Commercial Approaches to Genetic Selection for Growth and Feed Conversion in Domestic Poultry.' *Poultry Science* 76 (1997): 1121–5.

Eriksson, J., et al., 'Identification of the *Yellow Skin* Gene Reveals a Hybrid Origin of the Domestic Chicken.' *PLoS Genetics* 4 (2008): 1–8.

Fairfull, R.W., et al. 'Poultry Breeding: Progress and Prospects for Genetic Improvement of Egg and Meat Production.' www.cgil.uoguelph.ca/pub/6wcgalp/6wcFairfull.pdf .

Felch, I.K. *The Breeding and Management of Poultry*. Hyde Park, NY: Norfolk County Press, 1877.

– *Poultry Culture: How to Raise, Mate and Judge Thoroughbred Fowls*. Chicago: Donohue, Henneberry, 1902.

Fisher, R.A. *The Theory of Inbreeding*. London: Oliver and Boyd, 1949.

Flock, D.K., and R. Preisinger. 'Commercial Breeding of Egg-Type Chickens to Maximize Egg Income over Feed Cost.' Forty-sixth Breeder Roundtable Proceedings, St Louis, MO, 1997.

– 'Genetic Changes in Layer Breeding: Historical Trends and Future Prospects.' Manuscript, ca. 1999.

– 'Specialization and Concentration as Contributing Factors to the Success of the Poultry Industry in the Global Food Market.' Draft for *Lohmann Information*, winter 2007.

Fraser, A. *Animal Husbandry Heresies*. London: Crosby Lockwood & Son Ltd, 1960.

Goto, E., and A.W. Nordskog. 'Heterosis in Poultry: Estimation of Combining

Ability Variance from Diallel Crosses of Inbred Lines in the Fowl.' *Poultry Science* 38 (1959): 1381–88.
Graham, W.R. 'Modern Poultry Tendencies.' *O.A.C. Review* 39 (1926): 55–6. http://jcgi.pathfinder.com/time/magazine/article/0,9171,846554,00.html.
Hagedoorn, A.L. *Animal Breeding*. London: Crosby Lockwood & Son Ltd, 1939; 6th ed., 1962.
Hagedoorn, A.L., and G. Skyes. *Poultry Breeding: Theory and Practice*. London: Crosby Lockwood & Son Ltd, 1953.
Hallauer, A.R. 'History, Contribution, and Future of Quantitative Genetics in Plant Breeding: Lessons from Maize.' *Crop Science* 47 (2007), from *International Plant Breeding Symposium*, 2007, S4–19.
Harper, M.H. *Breeding of Farm Animals*. New York: Orange Judd Co., 1920.
Harvey, W. *Disputations Touching the Generation of Animals* (1653). Repr. Oxford: Blackwell Scientific, 1981.
Havenstein, G.B. 'Performance Changes in Livestock Following 50 Years of Genetic Selection.' *Lohmann Information* 41 (December 2006): 30–7.
Havenstein, G.B., et al. 'Growth, Livability, and Feed Conversion of 1957 versus 2001 Broilers when Fed Representative 1957 and 2001 Broiler Diets.' *Poultry Science* 82 (2003): 1500–8.
Hays, F.A., and G.T. Klein. *Poultry Breeding Applied*. Mount Morris, IL: Watt Publishing Co., 1952.
Heisdorf, A. 'Twenty Years Experience with Reciprocal Recurrent Selection.' Eighteenth Annual Session, National Poultry Breeders' Roundtable, 1969, 112–19.
Hill, W.C., and T.F.C. Mackay, eds. *Evolution and Animal Breeding: Reviews on Molecular and Quantitative Approaches in Honour of Alan Robertson*. Wallingford, UK: C.A.B. International, 1989.
Housman, W. *The Improved Shorthorn, Notes and Reflections upon Some Facts in Shorthorn History, with Remarks upon Certain Principles of Breeding*. London: Ridgeway, 1876.
Hunton, P. '100 Years of Poultry Genetics.' *World's Poultry Science Journal* 62 (2006): 417–28.
Hutt, F.D. *Genetics of the Fowl* (1949). Blodgett, OR: Norton Creek Press, 2003.
– 'Seventy-five Years of Poultry Genetics.' Roundtable of Poultry Breeders Association, 1975. www.poultryscience.org/pba/1952–2003/1975/1975%20Hutt.pdf.
Hutt, F.D., and B.A. Rasmusen. *Animal Genetics*. 2nd ed. New York: John Wiley & Sons, (1964) 1982.
– 'Whither Poultry Genetics?' *Poultry Science* 21 (1965): 53–61.
Jasper, A.W. 'The Farmer and the Poultry Industry.' *World's Poultry Science Journal* 27 (1971): 43–9.

Johannsen, W. 'The Genotype Conception of Heredity.' *American Naturalist* 45 (1911): 129–59.
Jones, D.F. 'Dominance of Linked Factors as a Means of Accounting for Heterosis.' *Genetics* 2 (1917): 466–79.
– *Genetics in Plant and Animal Improvement.* New York: John Wiley & Sons, Inc., 1925.
Jull, M.A. 'Inbreeding and Crossbreeding in Poultry.' *Journal of Heredity* 24 (1933): 93–101.
– *Poultry Breeding.* 3rd ed. New York: John Wiley & Sons, Inc., 1952.
Kent, O.B., and H.D. Branion. *A Brief History of the Poultry Science Association and Its Journal, 1908–1958.* Ithaca: Poultry Science Association, 1958.
Kinney, T.B., Jr, 'Poultry Breeding Research in North America.' *World's Poultry Science Journal* 30–1 (1974–5): 8–27.
Laughlin, K. 'The Evolution of Genetics, Breeding, and Production.' Temperton Fellowship Report no. 15, Harper Adams University College, UK, June 2007.
Laurie, D.F. 'Poultry Breeding in South Australia.' *American Breeders' Magazine* 1 (1911): 52–60.
Lerner, I.M. *The Genetic Basis of Selection.* New York: John Wiley & Sons, Inc., 1958.
– 'Lethal and Sublethal Characters in Farm Animals.' *Journal of Heredity* 35 (1947): 219–24.
– *Population Genetics and Animal Improvement.* Cambridge: Cambridge University Press, 1950.
Lerner, I.M., and E.R. Dempster. 'Some Aspects of Evolutionary Theory in the Light of Recent Work on Animal Breeding.' *Evolution* 2 (1948): 19–28.
Lerner, I.M., and H. Donald. *Modern Developments in Animal Breeding.* London and New York: Academic Press, 1966.
Lerner, I.M., and L.N. Hazel. 'Population Genetics of a Poultry Flock under Artificial Selection.' *Genetics* 32 (1947): 325–39.
Lush, J. *Animal Breeding Plans.* Ames, IA: Collegiate Press, Inc., 1937
– 'Genetics and Animal Breeding.' In *Genetics in the 20th Century*, ed. L.C. Dunn, 493–525. New York: MacMillan Co., 1951.
– 'Improving One Character by Breeding for Another.' Fact Finding Conference of the Institute of American Poultry Industries, 1958–9, 109–38.
Mackay, T.F.C. 'Alan Robertson (1920–1989).' *Genetics* 125 (1990): 1–7.
Marks, H.L. 'Evaluation of the Athens-Canadian Randombred Population. 2. Comparison with Parental Population.' *Poultry Science* 50 (1971): 1507–9.
Marks, H.L., and P.B. Siegel. 'Evaluation of the Athens Canadian Randombred Population. 1. Time Trends at Two Locations.' *Poultry Science* 50 (1971): 1405–11.
McClary, C.F. 'Opening Remarks.' Ninth Annual Session, National Poultry Breeders' Roundtable, 1960, 9–10.

- 'Reciprocal Recurrent Selection Response in Poultry, Other Animals and Plants.' Eighteenth Annual Session, National Poultry Breeders' Roundtable, 1969, 120–49.
Merritt, E.S., and R.S. Gowe. 'Development and Genetic Properties of a Control Strain of Meat-type Fowl.' *Twelfth World's Poultry Conference Section Papers*, 66–70.
Mingay, E., ed. *Arthur Young and His Times*. London: Macmillan Press Ltd, 1975.
Morse, G.B. 'Poultry Pathology: Its Place in the Curriculum.' *O.A.C. Review* 22 (1910): 246–51.
New Rules Project. *The Agriculture Sector*, 'Canadian Supply Management System.' http://www.newrules.org/agri/CanadaSM.html.
- 'Supply Management.' http://www.newrules.org/agri/splymg2.html.
- 'US Supply Management Proposals.' http://www.newrules.org/agri/USsplmgmt.html.
- 'When the Farmer Makes the Rules.' http://www.newrules.org/journal/nrfall100farmer.html.
Nordskog, A.W. 'The Evolution of Animal Breeding Practices – Commercial and Experimental.' Fourteenth Annual Session, National Poultry Breeders' Roundtable, 1965, 51–74.
Nordskog, A.W., L.T. Smith, and R.E. Philips. 'Heterosis in Poultry: Crossbreds versus Top-Crosses.' *Poultry Science* 38 (1959): 1372–80.
Ontario Agricultural College. 'The Department of Poultry Husbandry.' *News Bulletin*, 1959.
Ontario Poultry Conference. *Proceedings*. 1957.
Orde, A., ed. *Mathew and George Culley: Travel Journals and Letters, 1765–1798*. Oxford: Oxford University Press, 2002.
Orozco, F., and J.L. Campo. 'A Comparison of Purebred and Crossbred Genetic Parameters in Layers.' *World's Poultry Science Journal* 30–1 (1974–5): 149–53.
Pearl, R. 'Inheritance of Hatching Quality of Eggs in Poultry.' *American Breeders' Magazine* 1 (1911): 129–33.
Poultry Breeders of America. *Fact Finding* (later *Proceedings*) of the Annual National Poultry Breeders' Roundtable, unpublished, 1958–69.
Prentice, E.P. 'Food for Americans, 1980– . The Supply of Animal Proteins: The Agricultural Colleges.' *Political Science Quarterly* 66 (1951): 481–506.
Reaman, G.E. *History of the Holstein-Friesian Breed in Canada*. Toronto: Collins, 1946.
Reliable Poultry Journal. *The Leghorns: Brown, White, Black, Buff and Duckwing*. Quincy, IL: Reliable Poultry Journal Publishing, 1901.
- *The Plymouth Rocks: Barred, White and Buff*. Quincy, IL: Reliable Poultry Journal Publishing Co., 1899.

Research Committee on Animal Breeding. 'Live-Stock Genetics.' *Journal of Heredity* 6 (1915): 21–31.

Rice, J. 'Report of the Committee on Breeding Poultry: Some Principles of Poultry Breeding.' *American Breeders Association*, 1909: 376–9.

van Riper, Walker. 'Aesthetic Notions in Animal Breeding.' *Quarterly Review of Biology* 7 (1932): 84–92.

Rishell, W.A. 'Breeding and Genetics. Symposium: Genetic Selection – Strategies for the Future, Breeding and Genetics – Historical Perspective.' *Poultry Science* 76 (1997): 1057–61.

Sanders, J.H. *The Breeds of Live Stock, and the Principles of Heredity.* Chicago: J.H. Sanders Publishing Co., 1887.

Scott, G. 'The Learning Revolution, Distance Learning and Its Application to a Global Poultry Industry: A UK Perspective.' Temperton Fellowship Report no. 11, Harper Adams University College, UK, 2003.

Sebright, Sir J.S. *The Art of Improving the Breeds of Domestic Animals.* London: John Harding, 1809.

'Shane Report: Canadian Industry at Crossroads.' *Watt Poultry.Com*, 2007. http://www.wattpoultry.com/Article.aspx?id=12148.

Shaw, T. *Animal Breeding.* Chicago: Orange Judd Co., 1901.

Shrader, H.L. 'The Chicken-of-Tomorrow Program: Its Influence on "Meat-Type" Poultry Production.' *Poultry Science* 31 (1952): 3–10.

Shull, G.H. 'What Is "Heterosis"?' *Genetics* 33 (1948): 439–46.

Siegel, J., et al. 'Progress from Chicken Genetics to the Chicken Genome.' *Poultry Science* 85 (2006): 2050–60.

Siegel, P.B., et al. 'Genetic Selection Strategies – Population Genetics.' *Poultry Science* 76 (1997): 1062–5.

Skinner, J.L. '150 Years of the Poultry Industry.' *World's Poultry Science Journal* 30–1 (1974–5): 27–31.

Slocum, R.R. 'Poultry Breeding.' *Journal of Heredity* 6 (1915): 483–7.

Snyder, E.S. 'A History of the Poultry Science Department at the Ontario Agricultural College, 1894–1968.' Unpublished manuscript, 1970.

Springer, N.M., and R.M. Stupar. 'Allelic Variation and Heterosis in Maize: How Do Two Haves Make More than a Whole?' *Genome Research* 17 (2007): 264–75.

Stevens, L. *Genetics and Evolution of the Domestic Fowl.* Cambridge: Cambridge University Press, 1991.

Termohlen, W.D. 'Past History and Future Developments.' *Poultry Science* 47 (1968): 6–22.

Troyer, A.F. 'Adaptedness and Heterosis in Corn and Mule Hybrids.' *Crop Science* 46 (2006): 528–43.

Van Tassell, C.P., et al. 'SNP Discovery and Allele Frequency Estimation by Deep Sequencing of Reduced Representation Libraries.' *Nature Methods* 5 (2008), 247–52.
Warren, D.C. *Practical Poultry Breeding*. New York: Macmillan Co., 1953.
– 'Techniques of Hybridization of Poultry.' *Poultry Science* 29 (1950): 59–63.
Wentworth, Lady (Judith Anne Blunt). *The Authentic Arabian Horse*. London: George Allen & Unwin Ltd, 1945.
– *Thoroughbred Racing Stock*. New York: Charles Scribner's Sons, 1938.
Wriedt, C. *Heredity in Live Stock*. London: Macmillan and Co., 1930.
Wright, L. *The Brahma Fowl: A Monograph*. London: Journal of Horticulture and Cottage Gardener, 1870.
Wright, S. *Evolution and the Genetics of Populations*, volume 3, *Experimental Results and Evolutionary Deductions*. Chicago: University of Chicago Press, 1977.
– 'Mendelian Analysis of the Pure Bred Breeds of Livestock, Part 1: The Measurement of Inbreeding and Relationship.' *Journal of Heredity* 14 (1923): 339–48.
– 'Mendelian Analysis of the Pure Bred Breeds of Livestock, Part 2: The Duchess Family of Shorthorns as Bred by Thomas Bates.' *Journal of Heredity* 14 (1923): 405–22.
– 'The Relation of Livestock Breeding to Theories of Evolution.' *Journal of Animal Science* 46 (1978): 1192–1200.
– *Systems of Mating and Other Papers*, Ames: Iowa State College Press, 1958. Repr. of 'Systems of Mating,' *Genetics* 6 (1921): 111–78; 'Evolution in Mendelian Populations,' *Genetics* 16 (1931): 97–159; 'Correlation and Causation,' *Journal of Agricultural Research* 20 (1921): 557–85; and 'The Method of Path Coefficients,' *Annals of Mathematic Statistics* 5 (1934): 161–215.

Secondary Sources

BOOKS

Bowler, P.J. *The Eclipse of Darwinism, Anti-Darwinism Evolutionary Theories in the Decades around 1900*. Baltimore: Johns Hopkins University Press, 1983.
– *The History of an Idea*. Berkeley: University of California, 1984.
– *The Non-Darwinian Revolution: Reinterpreting a Historical Myth*. Baltimore: Johns Hopkins University Press, 1988.
Chapman, A.B., ed. *General and Quantitative Genetics*. Amsterdam: Elsevier Science Publishers, 1985.
Coleman, W., and C. Limoges, eds. *Studies in the History of Biology*. Baltimore: Johns Hopkins University Press, 1979.

Comstock, R.E. *Quantitative Genetics with Special Reference to Plant and Animal Breeding*. Ames: Iowa State University Press, 1996.
Crow, J.F. *Basic Concepts in Population, Quantitative, and Evolutionary Genetics*. New York: W.H. Freeman and Co., 1986.
Cunningham, E.P. *Quantitative Genetic Theory and Livestock Improvement*. Hanover, NH: University Press of New England, 1979.
Derry, M.E. *Bred for Perfection: Shorthorn Cattle, Collies, and Arabian Horses since 1800*. Baltimore: Johns Hopkins University Press, 2003.
– *Horses in Society: A Story of Breeding and Marketing Culture, 1800–1920*. Toronto: University of Toronto Press, 2006.
– *Ontario's Cattle Kingdom: Purebred Breeders and Their World, 1870–1920*. Toronto: University of Toronto Press, 2001.
Dreyer, P. *A Gardener Touched with Genius: The Life of Luther Burbank*. Los Angeles: University of California Press, 1985.
Dunn, L.C. *A Short History of Genetics: The Development of Some of the Main Lines of Thought, 1864–1939*. New York: McGraw-Hill, 1965.
Falconer, D.S. *Introduction to Quantitative Genetics*. Edinburgh: Oliver and Boyd, 1960.
Fitzgerald, D. *The Business of Breeding: Hybrid Corn in Illinois, 1890–1940*. Ithaca: Cornell University Press, 1990.
– *Every Farm a Factory*. New Haven: Yale University Press, 2003.
Freeman, S., and J.C. Herron. *Evolutionary Analysis*. Upper Saddle River, NJ: Prentice Hall, 1998.
Frolov, I.T. *Philosophy and History of Genetics: The Inquiry and the Debates*. London: Macdonald, 1991.
Funk, E.M. *Hatchery Operation and Management*. New York: John Wiley & Sons, Inc., 1955.
Futuyma, D.F. *Evolutionary Biology*. 3rd ed. Sunderland, MA: Sinauer Associates, Inc., 1998.
Glass, B., O. Temkin, and W.L. Straus, Jr, eds. *Forerunners of Darwin, 1745–1859*. Baltimore: Johns Hopkins University Press, 1951.
Goodall, D. *A History of Horse Breeding*. London: Robert Hall, 1977.
Green, E.L. *Genetics and Probability in Animal Breeding Experiments*. London: MacMillan, 1981.
Green-Armytage, S. *Extraordinary Chickens*. New York: Harry N. Abrams, Inc., 2000.
Hanke, O.A., ed. *American Poultry History, 1823–1973*. Madison, WI: American Poultry History Society, 1974.
Harman, O., and M.R. Dietrich, eds. *Rebels, Mavericks, and Heretics in Biology*. New Haven: Yale University Press, 2008.

Hartl, D.L. *A Primer of Population Genetics.* Sunderland, MA: Sinauer Associates, Inc., 1981.
Harwood, J. *Styles of Scientific Thought: The German Genetics Community, 1900–1933.* Chicago: University of Chicago Press, 1992.
– *Technology's Dilemma: Agricultural Colleges between Science and Practice in Germany, 1860–1934.* New York: Peter Lang Publishing Group, 2005.
Heaman, E.A. *The Inglorious Acts of Peace: Exhibitions in Canadian Society during the Nineteenth Century.* Toronto: University of Toronto Press, 1999.
Heath-Agnew, E. *A History of Hereford Cattle and Their Breeders.* London: Duckworth & Co. Ltd, 1983.
Johnson, P.C. *Farm Animals in the Making of America.* Des Moines, IA: Wallace Homestead Book Co., 1975.
Jones, L.A. *Mama Learned Us to Work.* Chapel Hill: University of North Carolina Press, 2002.
Kitcher, P. *In Mendel's Mirror.* Oxford: Oxford University Press, 2003.
Kloppenburg, J.R. *First the Seed: The Political Economy of Plant Biotechnology, 1492–2000.* Cambridge: Cambridge University Press, 1988.
Lawr, D. 'Development of Agricultural Education in Ontario, 1870–1910.' PhD thesis, University of Toronto, 1972.
Leeson, S. *The Ontario Poultry Industry: An Illustrated History.* Ontario Poultry Council, 1986.
Leeson, S., and J.D. Summers. *Broiler Breeder Production.* Nottingham: Nottingham University Press, 2000; repr. Guelph, ON: University Books, 2009.
Marcus, A.I. *Agricultural Science and the Quest for Legitimacy: Farmers, Agricultural Colleges, and Experiment Stations, 1870–1890.* Ames: Iowa State University Press, 1985.
Mayr, E., and W.B. Provine. *The Evolutionary Synthesis: Perspectives on the Unification of Biology.* Cambridge, MA: Harvard University Press, 1980.
Mazumdar, P.M.H. *Eugenics, Human Genetics and Human Failings: The Eugenics Society, Its Sources and Critics in Britain.* London: Routledge, 1992.
– *Species and Specificity: An Interpretation of the History of Immunology.* Cambridge: Cambridge University Press, 1995.
McCormick, V. *Farm Wife: A Self-Portrait, 1886–1896.* Ames: Iowa State University Press, 1990.
McMurry, S. *Transforming Rural Life: Dairying Families and Agricultural Change, 1820–1885.* Baltimore: Johns Hopkins University Press, 1995.
Montcrieff, E., S. Joseph, and I. Joseph. *Farm Animal Portraits.* Woodbridge, UK: Antique Collectors' Club, 1996.
Olby, R. *Origins of Mendelism.* 2nd ed. Chicago: University of Chicago Press, 1985.

Olson, A., and J. Voss. *The Organization of Knowledge in America, 1860–1920*. Baltimore: Johns Hopkins University Press, 1979.
Pauly, P.J. *Controlling Life: Jacques Loeb and the Engineering Ideal of Biology*. New York: Oxford University Press, 1987.
Pawson, H.C. *Robert Bakewell: Pioneer Livestock Breeder*. London: Crosby Lockwood, 1957.
Persell, S.M. *Neo-Lamarckism and the Evolution Controversy in France, 1870–1920*. Lewiston: Edwin Mellen Press, 1999.
Pirchner, F. *Population Genetics in Animal Breeding*. San Francisco: W.H. Freeman and Co., 1969.
Pollak, E., et al., eds. *Proceedings of the International Conference on Quantitative Genetics*. Ames: Iowa State University Press, 1977.
Powell, F.W. *Bureau of Animal Industry: Its History, and Organization*. New York: AMS Press, 1974.
Provine, W.B. *The Origins of Theoretical Population Genetics*. Chicago: University of Chicago Press, 1971.
– *Sewell Wright and Evolutionary Biology*. Chicago: University of Chicago Press, 1986.
Quirk, L. *Prof. William Richard Graham: Poultryman of the Century*. Guelph: University of Guelph, 2005.
Rader, K. *Making Mice: Standardizing Animals for American Biomedical Research, 1900–1955*. Princeton: Princeton University Press, 2004.
Ritvo, H. *The Animal Estate*. Cambridge, MA: Harvard University Press, 1987.
– *The English and Other Creatures in the Victorian Age*. Cambridge, MA: Harvard University Press, 1987.
– *The Platypus and the Mermaid and Other Figments of the Classifying Imagination*. Cambridge, MA: Harvard University Press, 1997.
Rogers, E.M. *Diffusion of Innovations*. New York: Free Press of Glencoe, 1962; 4th ed., 1995.
Ross, A.M., and T.A. Crowley. *The College on the Hill: A New History of the Ontario Agricultural College, 1874–1999*. 2nd ed. Toronto: Ontario Agricultural College Alumni Association and Dundurn Press, 1999.
Rossiter, M. *The Emergence of Agricultural Science: Justus Liebig and the Americans, 1840–1880*. New Haven: Yale University Press, 1975.
Rothschild, M., and S. Newman, eds. *Intellectual Property Rights in Animal Breeding and Genetics*. New York: CABI Publishing, 2002.
Russell, N. *Like Engend'ring Like: Heredity and Animal Breeding in Early Modern England*. Cambridge: Cambridge University Press, 1986.
Ryder, M.L. *Sheep and Man*. London: Duckworth, 1983.

Sapp, J. *Beyond the Gene: Cytoplasmic Inheritance and the Struggle for Authority in Genetics*. Oxford: Oxford University Press, 1987.
Sawyer, G. *The Agribusiness Poultry Industry: A History of Its Development*. New York: Exposition Press, 1971.
Schapsmeier, E.L., and F.H. Schapsmeier. *Henry A. Wallace of Iowa: The Agrarian Years, 1910–1940*. Ames: Iowa State University Press, 1968.
Scott, R. *The Reluctant Farmer: The Rise of Agricultural Extension to 1914*. Chicago: University of Chicago Press, 1970.
Scott, V.R. *Railway Development Programs in the Twentieth Century*. Ames: Iowa State University Press, 1985.
Serafini, A. *The Epic History of Biology*. New York: Plenum Press, 1993.
Smith, D.C. *The Maine Agricultural Experiment Station*. Orono: University of Maine, 1980.
Smocovitis, V.B. *Unifying Biology: The Evolutionary Synthesis and Evolutionary Biology*. Princeton: Princeton University Press, 1996.
Striffler, S. *Chicken: The Dangerous Transformation of America's Favorite Food*. New Haven: Yale University Press, 2005.
Strom, C. *Profiting from the Plains: The Great Northern Railway and Corporate Development of the American West*. Seattle: University of Washington Press, 2003.
Talbot, R.B. *The Chicken War: An International Trade Conflict between the United States and the European Economic Community, 1961–64*. Ames: Iowa State University Press, 1978.
Thomas, K. *Man and the Natural World: Changing Attitudes in England 1500–1800*. London: Allen Lane, 1983.
Thompson, J.A. *History of Livestock Raising in the United States, 1607–1860*. Washington: Bureau of Agricultural Economics, 1942.
Tozer, B. *The Horse in History*. London: Methuen, 1908.
Trow-Smith, R. *A History of British Livestock Husbandry, 1700–1900*. London: Routledge and Kegan Paul, 1959
Tudge, C. *In Mendel's Footnotes: An Introduction to the Science and Technologies of Genes and Genetics from the Nineteenth Century to the Twenty-Second*. London: Jonathan Cape, 2000.
Urquhart, M.C., and K.A. Buckley. *Historical Statistics of Canada*. Toronto: MacMillan Co. Ltd, 1965.
Weir, B.S., et al., eds. *Proceedings of the Second International Conference on Quantitative Genetics*. Sunderland, MA: Sinauer Associates, Inc., 1988.
Wood, R.J., and V. Orel. *Genetic Prehistory in Selective Breeding: A Prelude to Mendel*. Oxford: Oxford University Press, 2001.

ARTICLES

Alter, S.G. 'The Advantages of Obscurity: Charles Darwin's Negative Inference from the History of Domestic Breeds.' *Annals of Science* 64 (2007): 235–50.

Amidon, K.S. 'The Visible Hand and the New American Biology: Towards an Integrated Historiography of Railway-supported Agricultural Research.' *Agricultural History* 83 (2008): 309–36.

Beckett, J.V. Note on *Matthew and George Culley: Travel Journals and Letters, 1765–1798* in *English Historical Review* 118 (2003): 803–4.

Bonneuil, C. 'Mendelism, Plant Breeding and Experimental Cultures: Agriculture and the Development of Genetics in France.' *Journal of the History of Biology* 39 (2006): 281–308.

Bowie, G.G.S. 'New Sheep for Old: Changes in Sheep Farming in Hampshire, 1792–1879.' *Agricultural History Review* 35 (1987): 15–23.

Bowler, P.J. 'The Changing Meaning of "Evolution."' *Journal of the History of Ideas* 36 (1975): 95–104.

Bugos, G.E. 'Intellectual Property Protection in the American Chicken-breeding Industry.' *Business History Review* 66 (1992): 127–68.

Cain, J., and I. Layland. 'The Situation in Genetics: Dunn's 1927 Russian Tour.' *The Mendel Newspaper*, n.s., 12 (2003).

Carlson, L. 'Forging His Own Path: William Jasper Spillman and Progressive Era Breeding and Genetics.' *Agricultural History* 79 (2005): 50–73.

Castonguay, S. 'The Transformation of Agricultural Research in France: The Introduction of the American System.' *Minerva* 43 (2005): 265–87.

Churchill, R.B. 'William Johannsen and the Genotype Concept.' *Journal of the History of Biology* 7 (1974): 5–30.

Clutton-Brock, J. 'Darwin and the Domestication of Animals.' *Biologist* 29 (1982): 72–6.

Cooke, K.J. 'Expertise, Book Farming, and Government Agriculture: The Origins of Agricultural Seed Certification in the United States.' *Agricultural History* 76 (2002): 524–45.

– 'From Science to Practice, or Practice to Science? Chickens and Eggs in Raymond Pearl's Agricultural Breeding Research, 1907–1916.' *Isis* 88 (1997): 62–86.

Copus, A.K. 'Changing Markets and the Development of Sheep Breeds in Southern England, 1750–1900.' *Agricultural History Review* 37 (1999): 238–51.

Cornell, J.F. 'Analogy and Technology in Darwin's Vision of Nature.' *Journal of the History of Biology* 17 (1984): 202–344.

Crow, J.F. 'Sewell Wright's Place in Twentieth-Century Biology.' *Journal of the History of Biology* 23 (1990): 57–89.

Davis, B.D. 'The Background: Classical to Molecular Genetics.' In *The Genetic Revolution: Scientific Prospects and Public Perceptions*, ed. B.D. Davis. Baltimore: Johns Hopkins University Press, 1991.

Denison, R.F., E.T. Kiers, and S.A. West. 'Darwinian Agriculture: When Can Humans Find Solutions beyond the Reach of Natural Selection?' *Quarterly Journal of Biology* 78 (2003): 145–68.

Derry, M.E. 'Gender Conflicts in Dairying: Ontario's Butter Industry, 1880–1920.' *Ontario History* 90 (1998): 31–47.

Dunn, L.C. 'Genetics at the Anikowa Station: A Russian Animal Breeding Centre That Has Been Developed during the Reconstruction Period.' *Journal of Heredity* 19 (1928): 281–4.

– 'Poultry Genetics Up to Date.' *Journal of Heredity* 24 (1933): 198.

– 'The Transformation of Biology: A Geneticist's Viewpoint.' *Journal of Heredity* 57 (1966) 159–65.

Evans, L.T. 'Darwin's Use of the Analogy between Artificial and Natural Selection.' *Journal of the History of Biology* 17 (1984): 113–40.

Falconer, D.S. 'Early Selection Experiments.' *Annual Review of Genetics* 26 (1992): 1–14.

Fitzgerald, D. 'Beyond Tractors: The History of Technology in American Agriculture.' *Technology and Culture* 32 (1990): 114–26.

– 'Farmers Deskilled: Hybrid Corn and Farmers' Work.' *Technology and Culture* 34 (1993): 324–43.

Gisolfi, M.R. 'From Crop Lien to Contract Farming: The Roots of Agribusiness in the American South, 1929–1939.' *Agricultural History* 80 (2006).

Gliboff, S. 'Gregor Mendel and the Laws of Evolution.' *History of Science* 37 (1999): 217–35.

Grasseni, C. 'Managing Cows: An Ethnography of Breeding Practices and Uses of Reproductive Technology in Contemporary Dairy Farming in Lombardy (Italy).' *Studies in History and Philosophy of Biological and Biomedical Sciences* 38 (2007): 488–510.

Hartmann, W. 'From Mendel to Multi-national in Poultry Breeding.' *World's Poultry Science* 45 (1989): 5–26.

Harwood, J. 'Introduction to the Special Issue on Biology and Agriculture.' *Journal of the History of Biology* 39 (2006): 237–9.

– 'On the Genesis of Technoscience: A Case Study of German Agricultural Education.' *Perspectives on Science* 13 (2005): 329–51.

Horowitz, R. 'Making the Chicken of Tomorrow: Reworking Poultry as Commodities and as Creatures, 1945–1990.' In *Industrializing Organisms: Introducing Evolutionary History*, ed. S.R. Schrepfer and P. Scranton. London: Routledge, 2004.

Johnson, L.P.V. 'Dr. W.J. Spillman's Discoveries in Genetics: An Evaluation of His Pre-Mendelian Experiments with Wheat.' *Journal of Heredity* 39 (1948): 247–54.
Kevles, D.J. 'The Advent of Animal Patents: Innovation and Controversy in the Engineering and Ownership of Life,' in *Intellectual Property Rights and Patenting in Animal Breeding and Genetics*, ed. S. Newman and M. Rothschild (New York: CABI Publishing, 2002), 18–30.
– 'Patents, Protections, and Privileges.' *Isis* 98 (2007): 323–31.
– 'Protections, Privileges, and Patents: Intellectual Property in Animals and Plants since the Late Eighteenth Century.' In *Con/Texts of Invention*, ed. M. Biagioli, P. Jaszi, and M. Woodmansee (Chicago: University of Chicago Press, forthcoming).
Kimmelman, B. 'The American Breeders' Association: Genetics and Eugenics in an Agricultural Context, 1903–1913.' *Social Studies of Science* 13 (1983): 163–204.
– 'Mr. Blakeslee Builds His Dream House: Agricultural Institutions, Genetics, and Careers 1900–1945.' *Journal of the History of Biology* 39 (2006): 241–80.
Kingsland, S.E. 'The Battling Botanist: Daniel Trembly MacDougal, Mutation Theory, and the Rise of Experimental Evolutionary Biology in America, 1900–1912.' *Isis* 82 (1991): 479–509.
Kinney, T.B., Jr. 'Regional Quantitative Genetics Research on Poultry in the United States.' *World's Poultry Science* 24 (1968): 300–8.
Lerner, I.M. 'L.C. Dunn (1893–1974) and Poultry Genetics: A Brief Memoir.' *Journal of Heredity* 65 (1974): 185–6.
Loew, F.M. 'Animal Agriculture.' In *The Genetic Revolution: Scientific Prospects and Public Perceptions*, ed. B.D. Davis. Baltimore: Johns Hopkins University Press, 1991.
Marie, J. 'For Science, Love and Money: The Social Worlds of Poultry and Rabbit Breeding in Britain, 1900–1940.' *Social Studies of Science* 38 (2008): 919–36.
– 'The Situation in Genetics II: Dunn's 1927 European Tour.' *The Mendel Newspaper*, n.s., 13 (2004).
Mayr, E. 'The Nature of the Darwinian Revolution.' *Science* 176 (June 1972): 981–9.
McCormick, V. 'Butter and Egg Business: Implications from the Records of a Nineteenth-century Farm Wife.' *Ohio History* 100 (1991): 57–67.
Mendelsohn, J.A. '"Like All That Lives": Biology, Medicine and Bacteria in the Age of Pasteur and Koch.' *History and Philosophy of the Life Sciences* 24 (2002): 3–36.
Müller-Wille, S., and V. Orel. 'From Linnaean Species to Mendelian Factors: Elements of Hybridism, 1751–1870.' *Annals of Science* 64 (2007): 171–215.

Olby, R. 'Mendel No Mendelian?' *History of Science* 12 (1979): 53–72.
Orel, V. 'Cloning, Inbreeding, and History.' *Quarterly Review of Biology* 72 (1997): 437–40.
– 'Commemoration of the N.I. Vavilov Centennial at Brno.' *Folia Mendelianna* 23 (1988): 37–50.
– 'Selection Practice and Theory of Heredity in Moravia before Mendel.' *Folia Mendelianna* 12 (1977): 179–99.
– 'The Spectre of Inbreeding in the Early Investigation of Heredity.' *History and Philosophy of the Life Sciences* 19 (1997): 315–30.
Orel, V., and R.J. Wood. 'Early Development in Artificial Selection as a Background to Mendel's Research.' *History and Philosophy of the Life Sciences* 3 (1981): 145–70.
– 'Scientific Animal Breeding in Moravia before and after the Discovery of Mendel's Theory.' *Quarterly Review of Biology* 75 (2000): 149–57.
Palladino, P. 'Between Craft and Science: Plant Breeding, Mendelian Genetics, and British Universities, 1900–1920.' *Technology and Culture* 34 (1993): 300–23.
Pardue, S.L. 'Educational Opportunities and Challenges in Poultry Science: Impact of Resource Allocation and Industry Needs.' *Poultry Science* 76 (1997): 938–43.
Paul, D.B., and B. Kimmelman. 'Mendel in America: Theory and Practice, 1900–1919.' In *The American Development of Biology*, ed. R. Rainger, K. Benson, and J. Maienschein. Philadelphia: University of Pennsylvania Press, 1988.
Pauly, P.J. 'The Appearance of Academic Biology in Late Nineteenth-Century America.' *Journal of the History of Biology* 17 (1984): 369–97.
Pawson, H.C. 'Some Agricultural History Salvaged.' *Agricultural History Review* 7 (1959): 6–13.
Rheinberger, H.-J. and P. McLaughlin. 'Darwin's Experimental Natural History.' *Journal of the History of Biology* 17 (1984): 345–68.
Richards, R.A. 'Darwin and the Inefficacy of Artificial Selection.' *Studies in History and Philosophy of Science* 28 (1997): 75–97.
Riney-Kehrberg, P. 'Women, Technology, and Rural Life: Some Recent Literature.' *Technology and Culture* 38 (1997): 942–53.
Ritvo, H. 'Possessing Mother Nature: Genetic Capital in Eighteenth-Century Britain.' In *Early Modern Conceptions of Capital*, ed. J. Brewer and S. Staves, 413–26. London: Routledge, 1995.
Rowe, D.J. 'The Culleys, Northumberland Farmers 1767–1813.' *Agricultural History Review* 19 (1971): 156–74.
Ruse, M. 'Are Pictures Really Necessary? The Case of Sewell Wright's Adaptive Landscapes.' In *Picturing Knowledge: Historical and Philosophical Problems Concerning the Use of Art in Science*, ed. B.S. Baigrie. Toronto: University of Toronto Press, 1996.

- 'Charles Darwin and Artificial Selection.' *Journal of the History of Ideas* 36 (1975): 339–50.
Ryder, M.L. 'The History of Sheep Breeds in Britain.' *Agricultural History Review* 12 (1964): part 1, 1–12; part 2, 79–97.
Sapp, Jan. 'The Nine Lives of Gregor Mendel.' In *Experimental Inquiries: Historical, Philosophical and Social Studies of Experimentation in Science*, ed. H.E. Le Grand, 137–66. Dordrecht: Kluwer Academic Publishers, 1990.
- 'The Struggle for Authority in the Field of Heredity, 1900–1932: New Perspectives on the Rise of Genetics.' *Journal of the History of Biology* 16 (1983): 311–42.
Secord, J.A. 'Nature's Fancy: Charles Darwin and the Breeding of Pigeons.' *Isis* 72 (1981): 163–86.
Shaklee, W.E. 'Federal-Grant Funds and Poultry Breeding Research in the United States.' *World's Poultry Science* 29 (1972): 215–21.
- 'Regional Research on Poultry in the United States.' *World's Poultry Science* 24 (1968): 231–40.
Sloan, P.R. 'Essay Review: Ernst Mayr on the History of Biology.' *Journal of the History of Biology* 18 (1985): 145–53.
Stamhuis, I.H. 'A Female Contribution to Early Genetics: Tine Tammes and Mendel's Laws for Continuous Characters.' *Journal of the History of Biology* 28 (1995): 495–531.
Stamhuis, I.H., et al. 'Hugo de Vries on Heredity, 1889–1903: Statistics, Mendelian Laws, Pangenes, Mutations.' *Isis* 90 (1999): 238–67.
Sterrett, S.G. 'Darwin's Analogy between Artificial and Natural Selection: How Does It Go?' *Studies in History and Philosophy of Biological and Biomedical Sciences* 33 (2002): 151–68.
Termohlen, W.D. 'The History of Development of Poultry Departments in the State Colleges or Universities of the United States.' *Poultry Science* 46 (1967): 294–304.
Thaxton, Y.V., et al. 'The Decline of Academic Poultry Science in the United States of America.' *World's Poultry Science* 59 (2003): 303–13.
Theunissen, T. 'Breeding without Mendelism: Theory and Practice of Dairy Cattle Breeding in the Netherlands, 1900–1950.' *Journal of the History of Biology* 41 (2008): 637–76.
- 'Closing the Door on Hugo de Vries' Mendelism.' *Annals of Science* 51 (1994): 225–48.
- 'Darwin and His Pigeons: The Analogy between Artificial and Natural Selection Revisited,' *Journal of the History of Biology* 44 (2011); online as of October 2011, but not yet printed.
- 'The Holsteinization of the Friesians: Culture of Dairy Cattle Breeding in the Netherlands, 1945–1995.' *Isis*, forthcoming.

- 'Knowledge Is Power: Hugo de Vries on Science, Heredity and Social Progress.' *British Journal for the History of Science* 27 (1994): 291–331.
Thurtle, P. 'Harnessing Heredity in Gilded Age America: Middle Class Mores and Industrial Breeding in a Cultural Context.' *Journal of the History of Biology* 35 (2002): 43–78.
Warren, D.C. 'A Half Century of Advances in the Genetics and Breeding Improvement of Poultry.' *Poultry Science* 37 (1958): 3–20.
Wieland, T. 'Scientific Theory and Agricultural Practice: Plant Breeding in Germany from the late 19th to the Early 20th Century.' *Journal of the History of Biology* 39 (2006): 309–43.
Wilmot, S. 'Between the Farm and the Clinic: Agricultural and Reproductive Technology in the Twentieth Century.' *Studies in History and Philosophy of Biological and Biomedical Sciences* 38 (2007): 303–15.
- 'From "Public Service" to Artificial Insemination: Animal Breeding Science and Reproductive Research in Early 20th Century Britain.' *Studies in History and Philosophy of Biological and Biomedical Sciences* 38 (2007): 411–41.
Wood, R.J. 'Robert Bakewell (1725–1795), Pioneer Animal Breeder and His Influence on Charles Darwin.' *Folia Mendelianna* 8 (1973): 231–42.
Wood, R.J., and V. Orel, 'Scientific Breeding in Central Europe during the Early Nineteenth Century: Background to Mendel's Later Work.' *Journal of the History of Biology* 38 (2005): 239–72.
Wykes, D. 'Robert Bakewell (1725–1795) of Dishley: Farmer and Livestock Improver.' *Agricultural History Review* 52 (2004): 38–55.

OTHER

Access Aviagen. www.aviagen.com.
'Advent International: News Article,' at http://adventinternational.fr/News/Article.aspx?PageID=7.1&Ne.
Agriculture Canada. 'Poultry Marketplace – Overview of the Primary Poultry Breeder Industry.' http://www.agr.ca/poultry/brov-elap_e.htm.
- 'Primary Poultry Breeders: Company Profiles.' http://www.agr.gc.ca/poultry/brpr-elpr_m.htm.
Allard, R.W. 'Israel Michael Lerner, May 14, 1910–June 12, 1977.' National Academies Press Biographical Memoir. National Academy of Sciences. http://books.nap.edu/html/biomems/ilerner.html.
American Poultry Association. http://www.amerpoultryassn.com/poultryfancy.htm.
'B.C. Partners – About Us,' at http://www.bcpartners.com/bcp/aboutus; and 'B.C. Partners – Case Studies,' at http://bcpartners.com/bcp/cases/.

The Chicken: Its Biological, Social, Cultural, and Industrial History from Neolithic Middens to McNuggets. International Chicken Conference, Agrarian Studies, Yale University, May 2002. http://www.yale.edu/agrarianstudies/chicken/description.html.

'History of Breeds – Incubation and Embryology.' University of Illinois. http://www.urbanext.uiuc.edu/eggs/res10-breedhistory.html Accessed 11 September 2006.

Hy-Line International. http://www.hyline.com.

'The Impact of Genetics on Breeder Management.' From *Hybro B.V.*, May 2005; reproduced on *The Poultry Site*, http://www.thepoultrysite.com/FeaturedArticle/FAType.asp?AREA.

'Sequencing of the Chicken Genome: An Overview.' From *Nature*, December 2004; reproduced on *The Poultry Site*, http://www.thepoultrysite.com/FeaturedArticle/FAType.asp?AREA.

United Poultry Concerns, Inc. http://www.upc-online.org.

University of California. 'In Memoriam, September 1978 – I. Michael Lerner, Genetics: Berkeley.' http://content.cdlib.org/view?docId=hb4q2nb2nd&chunk.id=div00043.

University of Maryland. 'Papers of Morley A. Jull.' http://lib.umd.edu/archivesum/actions.DisplayEADDoc.do;jsessionid.

Index

Africa, 12
Alberta, 39, 75
Allard, R.W., 175
American Association of Instructors and Investigators in Poultry Husbandry, 115
American Civil War, 98
American Poultry Association, 9, 37, 49, 62, 102, 107, 108, 109, 121–4, 131, 142, 163, 204
Angus cattle, 196, 197
Anikowo, Russia, 87
Appleton, J.M., 171–2
Arabian horse, 23–8
artificial insemination, 185, 185, 212
Asmundson, V.F., 113, 123
Atlantic and Pacific Tea Company (A & P Food Stores), 165
Austria, 131

Bacon, Francis, 75
Bakewell, Robert, 7, 16–20, 21–2, 23, 25, 77, 90, 101, 106, 158, 185, 203, 211
Bates, Thomas, 25
Bateson, W., 81–2
Beal, W.J., 76
Belgium, 179

Bennett, J.C., 34
biological locks and breeding, 19, 21, 37, 45–9, 65–6, 83–5, 88, 90, 94, 95–6, 126, 142–6, 149, 151, 152, 163, 166, 174–5, 182–3, 186–8, 192, 193–5, 196–7, 209, 211–13
biometry, 91, 100
Boston, 59, 129
Brahma chicken breed, 31, 34, 77
Bray, F., 141
Breeders' Trotting Stud Book, 27
breeding, 16–20, 20–2, 23–8, 64, 67–8, 88–9, 93, 95, 96, 107–8; and organization of, 23–8, 101, 107–8, 109, 111–18, 121, 122, 203–4, 206; and pedigrees, 24–7, 29, 31–2, 36–7, 39, 66–7, 94, 114–15
Britain, 12–13, 15, 16–20, 26, 29, 32, 33, 36, 76, 87, 98, 99, 112, 128–9, 131, 140, 147, 158, 169
British Columbia, 75, 113–14, 123, 137, 141
broiler/meat marketing, 65–6, 155–6, 163, 166–72, 179–80, 194, 198–201, 208
Brown, E., 129
Brown, W., 77

Buffon, G.L.L., 76
Burnham, George, 29, 31, 37

Caesar, Julius, 12
California, 123, 124, 129
Canada, 33, 40, 57, 59, 77, 86, 95, 97–9, 109, 131, 180; and management supply, 6, 198–201
Canadian Baby Chick Association, 132, 135, 136
Canadian Chicken Marketing Agency (Chicken Farmers of Canada), 198
Canadian Randombred Control (ACRBC), 185–6
Carefoot, W.C., 189
Castle, W., 88–9, 104
chicken breeding, 5, 11, 12, 13, 22, 34, 36, 77–8, 78, 85–8, 107–8; and art and science, 4–5, 101–8, 111, 118, 120–3, 149, 150, 174–7, 201, 202, 205, 209–13; and art and science view variation between Canada and the United States, 121–5; and beauty, 188–9; and beauty/use dichotomy, 40–3, 45, 47–9, 108–9, 113–14, 119, 120, 121–5, 162, 204; and breeds (listed by breed); and breeding shape, 49–57; and corporate culture, 166, 186–8, 191–7, 209, 211, 212; and crossbreeding, 34, 36, 65–6, 77–8, 86, 145, 162–6; and disease, 193, 199; and domestication of *Gallus gallus*, 12; and double cross mating, 47–9, 65; and early breeding methodology, 13, 15, 17, 22, 33, 34–5, 36, 77–8, 81, 86–8; and egg breeding, 42–3, 59–64, 100–5, 125, 184–5, 190–1; and exhibition breeding, 40, 42–3, 45, 204; and Felch's chart, 59–60; and hybrid corn method, 94, 95–6, 143–4, 149–50, 174–5; 183–4; and inbreeding, 20–2, 59–64, 65–6, 67–8, 86, 89–90, 96, 182–8; and meat breeding, 42–3, 65–6, 77–8, 145, 156–8, 162–3, 166, 184–5, 173–4, 194–6; and Reciprocal Recurrent Selection, 183, 191; and specialization, 156–63, 173–4, 207–8; and sport cock fighting, 12, 13, 14, 15, 16, 20, 22, 29, 33; and standardbred breeding, 36–7, 39, 112, 113–14, 116; and Stoddard's system, 62–4; and women as breeders, 68–9, 156–62, 173–4, 207–8. *See also* Record of Performance (ROP) for poultry, in Canada; Record of Performance (ROP) for poultry, in the United States
chicken sexing, 140–2
Chicken-of-Tomorrow contest, 165–6, 182, 186, 187, 193, 194
Chicken War, 179
China, 12
Cleveland, C.D., 107
Clydesdale horse, 116
Coates, George, 25
Cochin chicken breed, 29, 30, 31, 34
cold storage, 97–9
Cole, L., 121
Cole, R.K., 191, 192, 193
Colling, Charles and Robert, 18, 25, 28
Connecticut, 34, 74, 85, 156
Connecticut Poultry Club, 37
consumption levels of poultry products, 97, 170–1, 179–80
Cornell University, 78, 123, 125, 191
Cornish chicken breed, 29, 34, 40, 41, 100, 162, 166, 181–2, 194

corporate breeding culture, 5–6, 19–20, 26, 143–4, 145–6, 149–51, 154, 166, 169–70, 174–7, 178, 180, 181–5, 186–8, 191–7, 208, 209, 211, 212; and education/research, 148–51
Crawford, R.D., 189
Culley, George, 16, 17, 19
Cushman, S., 77
Cyphers, C.C., 132

Darwin, C., and Darwinism, 7, 78–9, 86, 88, 91, 99, 175, 202
DeKalb family, 140, 145
Delaware, 95, 163, 165
Delmarva, 163, 169
Denmark, 87, 118, 131
Dishley Society, 19–20
Dominique chicken breed, 34
Donald, H., 175–7
Dorking chicken breed, 14, 34
drug companies and chicken breeding, 180–1, 208
Drumm, M., 138
Dryden, James, 119, 121
Dryden, John, 45
Dunn, L.C., 85, 86–8
Duston, A.D., 113

East, E.M., 211
Eaton, J.M., 78
egg industry, 57, 59, 125–6; and hen laying capacity, 126, 185–6; and marketing, 59, 71–2, 97–9, 144–6, 169–70, 180, 184, 194
Elford, F.C., 150
European Economic Community (EEU), 179

Farm Products Marketing Agencies Act, 198

feed companies and the chicken industry, 141–2, 166, 167–8, 170, 190, 196, 208
Felch, I.K., 54, 57, 59–62, 64, 68, 107, 108, 172, 191, 204, 205
Festetics, Count Imre, 76
First World War, 99, 112
Fishel, U.R., 35, 42–3, 44, 58
Fisher, R.A., 91, 92
Fisher, W.A., 129
flock size, 172–3
Ford, H., 119
Forsyth, R.M., 141
France, 7, 26, 76, 129, 131, 179
Fraser, A., 176–7
Frost, O.F., 34

Galloway cattle, 197
Game chicken breed, 14, 15, 29, 34, 40, 41, 42, 100
General Agreement on Tariffs and Trade (GATT), 179
General Stud Book (GSB), 24, 26
genetics, 7, 11, 74–96; and biometry, 91, 100; and chicken productivity, 185–6; and concepts, 81–3, 84, 92; and early chicken breeding experiments, 85–8, 99–101; and genomics, 8, 12, 178, 210; and hybrid corn breeding method (*see* plant breeding, and corn breeding; chicken breeding, and hybrid corn method); and inbreeding, 75–6, 80–4, 86, 88–91, 94, 96; and large animal breeding, 95, 176; and natural history, 75–6, 78–9, 92, 93, 124, 175, 202; and nature versus nurture, 185–6, 192–3, 199; and path coefficient, 89–91, 92, 94, 143, 205; and population genet-

ics, 91–5; quantitative genetics, 8, 91–5, 149–50, 178, 185–6, 211–21; and unit character, 80–1, 82. *See also* hybridizing and heterosis; Mendel and Mendelism
Georgia, 168, 169, 171, 186
Germany, 6, 17, 87, 118, 179
Gilbert, A.G., 77
Goodale, H.D., 85, 86, 120
government and agriculture in North America, 6, 74–5, 77–8, 85, 89, 93, 94–5, 97, 112–13, 115–18, 134–7, 138–9, 148–52, 180, 186, 198–201
Gowe, R.S., 150, 151, 193
Gowell, G.M., 119
Graham, W., 112, 113, 124, 125, 130, 135, 150, 160, 161, 172, 173
Graves, J., 129
Greece, 12, 131
Guernsey cattle, 120

Hadley, P.B., 85
Hagedoorn, A.L., 117, 123, 139, 176
Haldane, J.B.S., 91, 92
Haley, Alex, 15
Hannas, R.R, 137, 138
Hanson, J.A., 119, 121, 159
Hatch Act, 74
Hatchery Approval Policy, 136, 145
Hawkins, A.C., 54
Hen Fever, 40, 42
Herefords, 18, 196, 197
heterozygosis, 81, 83, 183
Hewes, T., 122, 123
Hicks, R.V., 139
Holstein Friesian cattle, 112, 159, 196
homozygosis, 81, 82, 83, 84, 89
House, C.A., 114
Hutt, F.B, 125, 149, 211

hybridizing and heterosis, 19, 21, 65–6, 75–6, 77, 78, 79, 80, 83–5, 86, 88–9, 90, 94, 95–6, 100, 106–7, 119, 126, 143–4, 163, 166, 174–5, 176, 178, 181–2, 183, 184, 190, 191, 196–7, 205, 209–11

Illinois, 57, 129
Illumina, Bovine SNP 50 Beadchip marker, 8
incubation technology, 128–30, 131–3, 205; and American/Canadian trade in baby chicks, 132, 133, 137–8, 142, 144–6
India, 12
Indiana, 57, 144
integration of the broiler industry, 166–74, 179, 180, 186, 198–201, 208
International Baby Chick Association, 131–2, 136, 137, 138, 139
Iowa, 93, 116, 118, 210
Iran, 12
Iroquois, 15
Italy, 12, 36, 131, 179

James I, King of England, 23
Japan, 140–1, 210
Java chicken breed, 34
Jerome, F.N., 150
Jersey, Lord, 26
Jersey cattle, 112, 159
Jewell, J.D., 168, 169
Johannsen, W., 81–2, 99, 205
Johnson, M., 118–19, 159
Jones, D.F., 83–4, 211
Jull, M.A., 98, 107, 125, 136

Kansas, 92, 94, 106, 136
Kimber, J., 120, 139, 159
Knight, T.A., 76

Kölreuter, J.G., 76

Lancashire, 36
Laurent, C.K., 171
Leghorn (Brown) chicken breed, 36
Leghorn (White) chicken breed, 33, 36, 54–7, 120, 121, 130, 140, 141, 143, 150, 151, 156, 157, 159, 184, 190
Leibig, Justin, 6
Leicester sheep, 16
Lerner, I.M., 124, 175–7
Lippincott, W.A., 85
Little, C.C., 89
Lloyd, E.A., 113, 123
London Poultry Club, 36
Longhorn cattle (English), 16, 158
Lush, J.L., 91–5, 149, 210
Luxembourg, 179

MacDougal, D.T., 79, 80
Maine, 9, 34, 77, 85, 102, 119
Malay chicken breed, 34
Manhattan Plan, 136–7, 139
Manitoba, 75, 125, 129
Marek's disease, 193, 199
Markham, Gervasse, 13
Martin, J.H., 145
Maryland, 125, 163, 165
Massachusetts, 34, 74, 85, 116, 120
Maupertius, Pierre-Louis de, 75
Medes, 12
Mendel, Gregor, and Mendelism, 7, 76, 78–80, 81–2, 85–6, 87, 88, 89, 91, 92, 93, 94, 96, 104, 105–8, 176, 204, 209
Mexico, 57, 133
Michigan, 76, 131
Miller, J., 117
Minnesota, 57, 118
Minorca chicken breed, 34, 36

Missouri, 119, 138
Monroe, S., 151
Montreal, 57, 72
Moravia, 7
Morgan, T.H., 99
Morman, J.B., 105
Morrell Land-Grant Act, 74
Moscow, 87
Mount Hope, 120, 190

National Trotting Association, 27
Neo-Lamarkism, 78
Netherlands, 7, 124, 176, 179
Neuhauser, E., 144
New Hampshire poultry breed, 163, 166
New Jersey, 116, 129
New York City, 37, 57, 59, 156
New York State, 15, 37, 42, 77, 107, 156, 190
Norman, C.A., 139
North America, 15, 16, 26, 36, 76, 77, 91
North Carolina, 71, 160
Norway, 87
Nova Scotia, 75

Ohio, 57, 72, 137, 138
Ontario, 42, 72, 129, 132, 135, 141, 144, 145, 159, 160; and Ontario Agricultural College, 9, 74, 77, 78, 112, 130, 145, 151, 172, 173; and Ontario Poultry Association, 42, 45
Oregon, 119
Ottawa, 112

Pearce, H.C., 165
Pearl, R., 85, 99–105, 119, 124, 149
Pearson, K., 91
Persia, 12

pigeon breeding, 17, 20, 21, 22, 33, 78, 107
plant breeding, 74–6, 78–80, 81, 83, 87, 138, 149, 181, 205; and corn breeding, 7, 76, 80, 83–5, 95–6, 143–4, 203, 209, 211
Plymouth Rock (Barred) chicken breed, 33, 34, 36, 37, 50–4, 71, 72, 77, 100, 122, 130, 146, 157, 160, 163, 184, 191; and breeding by double cross mating, 47–9
Plymouth Rock (White) chicken breed, 33, 34, 35, 42–3, 44, 54, 58, 166, 182, 191, 194
Poland, 87
Polish chicken breed, 45, 46
Portugal, 15, 194
Poultry Science Association, 124
Prentice, E.P., 120, 139, 159
producer/growers, 64–6, 129, 130–1, 133–4, 140–2, 145–7, 151, 152, 168, 193, 205–6, 207; and gender, 68–73, 173–4, 206, 207, 208
Punnett, R.C., 140

Quebec, 75, 129, 132

Reciprocal Recurrent Selection (RRS), 183, 191
Record of Performance (general), 117–18
Record of Performance for cattle, 112, 159
Record of Performance (ROP) for poultry, in Canada, 111–15, 122, 125, 135–6, 137, 139, 145, 146–7, 150, 151, 159, 190–1, 206–7
Record of Performance (ROP) for poultry, in the United States, 115–18, 125, 136–40, 146–7, 206–7

Rhode Island, 77–8, 85
Rhode Island Red chicken breed, 34, 160, 163, 164, 166, 182, 184, 194
Rice, J.E., 78, 124
Robertson, G., 150
Rome/Romans, 12, 13, 16, 36, 179, 209
Russia, 87, 99

Sachs, J. von, 79
Saglio, H., 166, 194
Saskatchewan, 75, 123
Saunders, William, 75
Sawyer, C.B., 139
science/craft attitudes to breeding, 5, 11, 88–9, 99–105, 106–9, 111, 118–25, 146–7, 174–7, 205–8
Scotland, 29, 87, 175, 177, 181, 187
Sebright, Sir John, 16, 20–2, 39, 60, 62, 89, 106, 203, 204, 205
Second World War, 99, 156, 169, 190
Seiling, A., 145
Serebrovsky, A.S., 87
Sewell, F.L., 54
Shanghai chicken breed, 29
Shaver, D. McQ., 171, 181, 189–97; and Leghorn Starcross 190, 191, 193, 199; and Shaver Blend cattle, 196–7
sheep breeding, 158
Shire horse, 16
Shorthorn cattle, 18, 25, 93, 112, 116, 158, 159, 197
Shull, G., 83–4, 191
Slocum, R.R., 101
SNP (single nucleotide polymorphism), 8
Snyder, E.S., 145
South America, 15
Spain, 15, 194
Spillman, W.J., 76

Spitzer, G.R., 139
Stahmer, L.A., 54
Standardbred horse, 27–8
Standard of Excellence/Perfection, 36–7, 38, 49, 65, 108, 112–14, 121–4, 137, 145, 162; and conflict over illustrations of, 49–57
started pullets, 142
Steele, Mrs Wilmer, 163
Stoddard, H.H., 62–4, 102–5, 106, 107, 119, 204, 205
Stothart, J., 150
Sweden, 131

Tancred, D., 119, 159
Texas, 118, 190
Thompson, E.B., 121–2
Thoroughbred horse, 23–8
Tomkins, Benjamin, 18
Toronto, 57, 72, 122, 125
trapnests, 100, 108–9, 111–12, 119, 121, 122–3

United States, 15, 26, 27–8, 29, 32, 33, 40, 57, 59, 77, 80, 86, 95, 97–9, 109, 131, 180
United States Poultry Improvement Plan (NPIP), 139
Upham, D.A., 34
Utility Poultry Club of England, 109

Vancouver, 113
Vantress, C., 165, 194
venture capital and chicken breeding, 181–2, 208–9
Virginia, 163
Vries, H. de, 79, 80, 176

Wallace, H.A. and H.B., 140, 143–4, 149
Warmblood horses, 26–7
Warren, D.C., 94, 106
Washington State, 76, 116
Watson, James D. and Francis Crick, 8
Weatherby, James, 24
West Virginia, 118, 125
Winnipeg, 129
Wisconsin, 118, 125, 210
women and the poultry industry, 68–73, 129, 130–1, 158–62, 173–4, 207–8
Wright, S., 93; and path coefficient, 89–91, 92, 94, 143, 205; and population genetics, 92
Wyandotte chicken breed, 160

Yorkshire, 36
Young, Arthur, 16, 20, 22